INDIVIDUAL ONSITE WASTEWATER SYSTEMS

Proceedings of the Third National Conference 1976

Sponsored by

National Sanitation Foundation

and

U.S. Environmental Protection Agency
Technology Transfer Program

Edited by

NINA I. McCLELLAND

Vice President, Technical Services
National Sanitation Foundation
Ann Arbor, Michigan

ANN ARBOR SCIENCE
PUBLISHERS INC
P.O. BOX 1425 • ANN ARBOR, MICH. 48106

Copyright © 1977 by Ann Arbor Science Publishers, Inc.
230 Collingwood, P. O. Box 1425, Ann Arbor, Michigan 48106

Library of Congress Catalog Card No. 76-050983
ISBN   0-250-40156-8

Manufactured in the United States of America
All Rights Reserved

PREFACE

The Third National Conference for Individual Onsite Wastewater Systems, cosponsored by NSF and the EPA Technology Transfer Program, was held in Ann Arbor, Michigan in November, 1976. Each conference to date has provided a forum for regulatory officials, equipment manufacturers, planners, land developers, and users to exchange current information relating to the state-of-the-art and field experiences with alternative systems for onlot treatment and disposal of wastewater. Identification of future research and demonstration needs has been a further expressed objective. These conferences are a corollary to NSF's current testing and listing program for Standard No. 40, relating to individual aerobic wastewater treatment plants, and a standard relating to wastewater recycle and water conservation devices, which is in the final stages of development. A fourth national conference has been scheduled for November, 1977 to assure continuing dialogue in this area so pertinent to achieving the environmental quality goals for which we strive.

Nina I. McClelland, 1977

ACKNOWLEDGMENTS

The Third National Conference on Individual Onsite Wastewater Systems was cosponsored by the U.S. Environmental Protection Agency Technology Transfer Program under Grant No. R805023-01. The National Sanitation Foundation gratefully acknowledges EPA's continued interest and support, and recognizes in particular the valuable assistance of Denis J. Lussier, who has served as Project Officer for the past two conferences.

The effective efforts of this year's planning committee are also greatly appreciated:

> Joe D. Brown, Director
> Division of Sanitary Engineering
> State Board of Health
> Jackson, Mississippi
> (Conference of State Sanitary Engineers)
>
> Dale Carlson, Dean
> College of Engineering
> University of Washington
> Seattle, Washington
>
> John Gannon
> School of Public Health
> University of Michigan
> Ann Arbor, Michigan
> (Task Committee on Health and Environmental Effects, Environmental Engineering Division, American Society of Civil Engineers)
>
> James F. Kreissl, Sanitary Engineer
> Municipal Environmental Research Laboratory
> Wastewater Research Division
> U.S. Environmental Protection Agency
> Cincinnati, Ohio

Orville Lee
U.S. Housing and Urban Development
Washington, D.C.

Denis J. Lussier
Office of Technology Transfer
U.S. Environmental Protection Agency
Cincinnati, Ohio

David MacLaren, President
Jet Aeration Company
Cleveland, Ohio
(Aerobic Industry)

Paul Pate, Director
Jefferson County Department of Health
Birmingham, Alabama
(Conference of Local Environmental Health Administrators)

Kenneth C. Pearson
Interlink Life Support Systems
Costa Mesa, California
(Recycle Industry)

Nicholas Pohlit, Executive Director
National Environmental Health Association
Denver, Colorado

Joseph A. Salvato, Jr.
New York State Department of Health
Albany, New York
(Great Lakes Upper Mississippi River Board)

Steve Serdahely, Regional Advisor
Water Supplies and Sewerage
c/o Pan American Health Organization/World
  Health Organization
Washington, D.C.

Lowell A. Welker, Director
Environmental Health
Allen County General Health District
Lima, Ohio

Robert Wheatley, Chief
Illinois State Health Department
Division of General Sanitation
Springfield, Illinois

Appreciation is also expressed to the authors, presiders, and registrants for participating in this year's conference, and to the dedicated NSF staff who finalized the program and completed local arrangements.

CONTENTS

1. The Next Steps in Protecting the Environment 1
   Senator Jennings Randolph
2. A Consultant's View on Waste Disposal 7
   Charles C. Johnson, Jr.
3. The Social, Economic and Political Impacts of Onlot Sewage Disposal 15
   Maurice K. Goddard
4. U.S. EPA Response to PL 92-500 Relating to Rural Wastewater Problems—Office of Research and Development (Sections 104 and 105) 21
   James F. Kreissl
5. U.S. EPA Response to PL 92-500 Relating to Rural Wastewater Problems—Office of Water Program Operations (Section 201) 37
   Keith Dearth
6. HUD's Response to the Housing Crisis—Current Extent of Agency Involvement 43
   Orville G. Lee
7. Alternative Methods of Regulating Onsite Domestic Sewerage Systems 53
   David E. Stewart
8. Onsite Wastewater Disposal: A Local Government Dilemma 67
   Paul Pate
9. An Overview of Disposal Options: The Ontario Program . . 75
   D. M. C. Saunders
10. The Need for Improving Septic System Repair Practices . . 87
    William L. Mellen
11. Field Application: Sand Mound and Evapotranspiration Systems 93
    Glenn E. Maurer
12. Effluent Quality Considerations Affecting the Use of Sand Filters and Oxidation Lagoons 103
    Michael Hines

13. Treatment Systems Required for Surface Discharge of
    Onsite Wastewater 113
    David K. Sauer
14. The Sewage Osmosis Concept for Onsite Disposal Systems—
    Clay Soils 131
    Frank P. Coolbroth
15. Soil Evaluation of Sites for Absorption Systems 139
    Dale E. Parker
16. Septage Disposal in Wastewater Treatment Plants 147
    Ivan A. Cooper and Joseph W. Rezek
17. Collection Alternative: The Pressure Sewer 171
    W. C. Bowne
18. Management Guidelines for Conventional and Alternative
    Onsite Sewage Systems—Washington State 187
    Gary Plews
19. Management Guidelines for Conventional and Alternative
    Onsite Sewage Systems—Pennsylvania 195
    William B. Middendorf
20. Report on the Ten State Committee for Onsite Sewage Systems 201
    Harold D. Baar
21. Individual Onsite Wastewater System Management in Colorado. 205
    Edmond B. Pugsley
22. New York State Standards for Individual Household Systems . 211
    Peter J. Smith
23. State Agency Management Plans and Approval Practices for
    Maine 215
    W. Clough Toppan
24. Integrated Waste Management Systems—Onsite MIUS
    Applications 221
    W. J. Boegly, Jr., W. R. Mixon and J. H. Rothenberg
25. The Boyd County Demonstration Project—A System
    Approach to Individual Rural Sanitation (An Update) . . . 235
    Lawrence E. Waldorf
26. Onsite Wastewater Facilities for Small Communities and
    Subdivisions 245
    Richard J. Otis
27. Integration of Onsite Disposal in a 201 Facilities Plan . . 277
    Jack L. Abney
28. Innovation in Wastewater Technology: The Challenge of
    the 1980s 301
    Kenneth C. Pearson
29. Pollution Control—Pathway to Perfection or Perdition . . . 307
    Dale A. Carlson
30. As We Now Look Ahead 311
    Robert M. Brown

KEYNOTE ADDRESS
THE NEXT STEPS IN PROTECTING THE ENVIRONMENT

The Honorable Jennings Randolph
>Senator of West Virginia
>United States Senate
>Public Works Committee
>Washington, D.C. 20510

It is a privilege to participate in this conference under the sponsorship of the National Sanitation Foundation. The desire of the American people to protect the environment in which we live has brought about a number of new governmental programs in recent years. Those of us with responsibility in the Congress have worked to develop and pass legislation that responds in a balanced fashion to the need to end pollution. We believe that we have moved in the right direction and, from time to time, we examine the impact of these programs and make necessary adjustments.

There has been progress toward our goals of ending pollution, but the success of this effort depends to a great degree on the people who actually implement the programs at state and local levels—people like the participants in this conference. Without you, there would be no success, and for this reason I value the opportunity to counsel with you.

The Senate Public Works Committee, which I have the responsibility to chair, has developed environmental laws over the past dozen years. These have enabled antipollution programs to be established. Committee members are now working on what might be called the second generation of environmental statutes. During the 94th Congress, we wrote new legislation relating to air and water pollution and solid waste management. Of these three major efforts, I regret that only the solid waste proposal, of which I was the principal sponsor, became law. The air and water pollution legislation were victims of a filibuster and a crowded calendar in the closing days of the session. It is our intention to address both of these areas during the next Congress which begins in early January.

The periodic reviews of environmental programs should not be interpreted as invitations to unduly weaken environmental laws. We remain firm in our belief that protection of the public health is an essential and desirable national objective. I believe also that environmental enhancement is consistent with sound economic growth. This is not a question of polarization.

These are times of change throughout our society, and the demands imposed by antipollution programs may seem particularly difficult in a time of economic recession. The difficulty faced by many industries may in part stem from the fact that there has been little or no consideration in the past for environmental protection. We must accept substantial expense of environmental controls as a cost of doing business and it need not be excessive. A recent study indicated that current environmental expenditures are only about 2% of our gross national product.

There are arguments that environmental requirements result in widespread unemployment. This assessment deserves close examination. The Environmental Protection Agency has reported that during the past five years, 75 plants closed and 15,710 workers lost their jobs because of antipollution laws. This loss is more than offset, however, by the new and rapidly growing environmental industry. The EPA reports that the same environmental laws have stimulated the establishment of an industry that now employs 1.1 million persons. This is the result of $15.7 billion in expected expenditures for pollution control and includes $4.2 billion spent by the federal government, $1.5 billion by state and local governments and $10 billion by private industry. The Council on Environmental Quality estimates that each billion-dollar expenditure generates an average of 70,000 jobs.

Pollution control, therefore, is one of our hottest growth industries with federal commitments of $200 billion expected during the 10-year period ending in 1983. Despite the substantial outlays, much work remains to be done in pollution control.

Your principal concern is with programs relating to water. Four years ago, in the Federal Water Pollution Control Act Amendments of 1972, the federal government established an ambitious program to restore the integrity of our country's water. We have long known that no society can exist without ample supplies of clean water. It is more recently that we have awakened to the fact that we as a nation have used our water carelessly. The program enacted in 1972 was intended to reverse the historic trends and to assure that our people would have the pure water they will need in the future. We knew then that the price of such an effort would be high. The benefits that would accrue, however, make it a reasonable expenditure.

The 1972 Act provided $18 billion in federal funds to help communities build pollution control facilities. The goals were ambitious and the enthusiasm was real. Yet, four years later, little more than half of the federal money available has been obligated. It is estimated that the 1977 deadlines for water pollution control will not be achieved in communities where 60% of our population lives.

This is cause for concern. Public Works Committee members will examine the water pollution program next year to determine both its strengths and its weaknesses. We want to know if we are moving in the right direction and if we have the proper tools to help meet the ambitious goals set in 1972. Those objectives themselves will be subjected to scrutiny to determine if they are valid in today's changed conditions and our expanded knowledge.

To assist us, we have four years of experience. We also have the findings and recommendations of the National Commission on Water Quality, authorized in 1972 to lay the groundwork for the type of mid-course correction we are beginning. I sponsored that portion of the 1972 Act authorizing the Commission study.

During the past year, it was the intention of the Senate Public Works Committee to address only the most urgent needs of the water pollution program. Among these were the construction grant program which required new authorizations. More than 20 states are virtually without federal funds to help build sewage treatment facilities. The House of Representatives chose to report and pass a broad-ranging comprehensive bill. As a result, it was **impossible** to reconcile our differences in the closing days of the 94th Congress.

I anticipate that the Senate Public Works Committee will move early in the new year to examine the critical issue of funding. We must decide whether to recommend a limited bill or to wait until we can propose legislation covering all water pollution issues. In addition to funding, we must assure that this money is distributed in a manner that will produce the greatest return. We must also address such questions as compliance deadlines, user charges, construction grant procedures and toxic substances. A major controversial issue is that concerned with permits for dredging and filling. This issue arose from judicial and administrative decisions relating to Section 404 of existing law. This was one of the controversial matters on which the Senate and House failed to agree this year.

I am sure that participants in this conference share the belief that money for operation of treatment plants should be increased. This is another matter we must examine. It is not enough merely to build treatment plants. They must be operated efficiently, effectively and safely. As more facilities are constructed, greater attention must be assigned to operational matters.

As a Senator from a state which is statistically the second most rural in the country, I am naturally concerned with providing sewage treatment services in lightly populated areas.

Despite the appropriation of billions of dollars, the 1970 census showed that 28.8% of the housing units in the United States are not served by public sewers. This means that 19.5 million households consisting of 58.5 million Americans must provide for themselves some method of home disposal for the nearly 3 billion gallons of domestic sewage which they generate daily. In the vast majority of cases, this situation creates serious health hazards.

The efforts initiated through federal grant programs to meet the need for proper collection, treatment and disposal of waste, which the environmental health of our nation demands, has been hampered by soaring construction costs. There is also a reluctance on the part of both federal and state agencies to adopt technological advances available to them in the area of sewage treatment. In many cases the preoccupation with centralized sewage systems has precipitated their construction where they are neither suited nor cost-effective. As a result, in many rural areas around the country, it is not uncommon to find the cost of conventional public wastewater collection and treatment facilities running between $5,000 to $10,000 per house.

One example of this problem is the Hepzibah Public Service District in Harrison County, West Virginia. This area does not have adequate public sanitary facilities and a new sanitation system serving 600 families is now under study. One segment of the proposed system which will serve 150 homes has an estimated construction cost of $1.2 million, or $8000 per family. Even with 75% grant funds from the Environmental Protection Agency, poorer rural communities cannot afford to finance such facilities.

Thus, it is impossible for local, state and federal governments to provide this most fundamental service to a large segment of the American people by conventional means. It is critical that alternative methods which have been developed to handle household sewage be given a responsible trial and that further development of them be aggressively pursued.

In my judgment, new technology has developed possible alternatives which make the treatment of waste for individual homes a practical, cost-effective way of meeting the problems of rural sanitation without the need for miles of costly collection lines and large treatment facilities.

Individual home treatment units could provide effective sanitation to millions of families living in areas which cannot be serviced economically by conventional methods. Such systems will reduce the cost of sanitation and help preserve our precious water supplies. Systems are available which can recycle treated, filtered and disinfected wastewater which can be reused for sanitary purposes. This technology could be of great importance in areas where scarce potable water should be reserved for cooking, drinking and bathing. Such programs can reduce by 40% the amount of water used by an average household.

One government agency, the Appalachian Regional Commission, has recognized the potential of alternative systems for serving more households at lower cost. The Commission is presently engaged in a demonstration project in Boyd County, Kentucky. A primary factor in the ARC project is that although most of the equipment in the demonstration serves single homes, all systems are owned, operated and maintained by a public sanitary district. This concept insures responsible maintenance and optimum operation.

Members of the Public Works Committee have expressed the view many times regarding the institutional uniqueness of the Appalachian Regional Commission as a federal/state partnership. I believe the Boyd County residential wastewater treatment demonstration illustrates an aspect of the Commission which has not often been commented upon: It is small enough to be able to respond to new ideas.

According to engineering studies, the estimated cost of providing centralized sewage facilities to the area of Boyd County served by the Appalachian demonstration project is approximately $565,000 or more than $9,000 per house.

The present cost of the demonstration is $200,000 or $4,200 per house. It should be noted, however, that this cost includes many items which are a result of the demonstration nature of the project. In addition, the innovative nature of the project resulted in rather inefficient installation procedures and higher costs. The lessons learned in Boyd County, however, will be of significant benefit in the efficient and economical development of future systems of this type. The Commission feels that, based on what they know from this demonstration, it is not unrealistic to expect the cost of future systems to be in the $2,500 to $3,000 range per house.

The effluent quality tests from the project indicate that the management concept developed for the project will permit optimum operation of the treatment. This is particularly demonstrated by the fact that equipment using surface discharge is meeting or exceeding standards of the Environmental Protection Agency.

I have urged the Environmental Protection Agency to fund such projects. Indeed, the law requires that such alternative systems be utilized where they can be shown to be cost-effective. I am informed that the EPA is now developing regulations for the utilization of systems to serve single homes.

This type of system is an example of the approach we should take to sanitation needs. There is no single method that will meet every need. We must be on the alert for new answers to the many questions we face.

Growing concern for the environment is a healthy sign as we complete our first two centuries as a nation. I am equally encouraged to see individuals in all walks of life express their interest in protecting the world which sustains us. I look forward to working with you in the future and to receiving your recommendations as we move toward our common goal—the building of a better America.

2
A CONSULTANT'S VIEW ON WASTE DISPOSAL

Charles C. Johnson, Jr.
    Resident Manager
    Malcolm Pirnie, Inc., Consulting Environmental Engineers
    Chairman of the National Drinking Water Advisory Council
    Washington, D.C. 20006

We must constantly remind ourselves that air, land and water comprise the environment with which we must be concerned. Of these, our water resources, more perhaps than any other, illustrate the interaction of all parts of the environment and particularly the recycling process that characterizes every resource of the ecosystem. Everything that man himself injects into the biosphere—chemical, biological or physical—can ultimately find its way into the earth's water. And these contaminants must be removed, by nature or by man, before that water is again potable. When we learn to protect our water resources, we move a long way toward the protection of the other media of the environment.

There are many things happening in our time that affect either the quality of the water we must treat for consumption and use, or affect the development of standards established to evaluate that quality. You are no doubt aware of some of these, but a few of particular interest are related below.

There is a national debate occurring at this time, accompanied by a court consideration, over what to do about organics in drinking water. One viewpoint recommends setting of a zero standard because of the widely accepted view of a no threshold level for known or suspected carcinogens. Another view says it is too early to establish a standard because available data do not allow an evaluation of the significance of trace amounts (ppb) of such substances in water. Nearly everyone agrees that the source of most of these organic compounds of significance to health originate with disposal of waste by our industries. The ultimate resolution of this problem can have far-reaching and significant impact on both the American public and the industrial sector.

If it could be determined that a threshold does exist for carcinogens, the establishment of a standard to protect the public health would be made considerably easier. The National Drinking Water Advisory Council has been pursuing this matter for a number of months now. Dr. Herbert Stokinger, a toxicologist with the National Institute of Occupational Safety and Health, presented a case for biologic thresholds at a recent meeting of the Council. Data were presented for several separate instances which indicated evidence that measurable and significant thresholds existed for eight suspected or known carcinogens. He noted that the carcinogens were of widely differing chemical structures, producing many different tumor types, presumably by different mechanisms. Further, he noted that thresholds were evident for all three major routes of entry and unrelated as to whether the carcinogens were of very high or low potency.

Dow Chemical Company has conducted studies on several substances and has obtained results similar to Stokinger's. In a study of hexachlorobutadiene, the Dow Chemical Company found that a high dosage caused multiple toxic effects, including kidney cancer in test rats, while lower doses did not.

Thresholds for health effects when discernible are essential for the development of sound environmental standards. In my opinion, it is incumbent on the scientific community to evaluate Dr. Stokinger's and the Dow Chemical Company's findings for the benefit of the public knowledge at the earliest possible time. Should it be determined that their work has merit and relevance, an entirely different approach becomes available for the establishment of effluent standards for many hazardous substances, and for the development of organic standards for water supplies.

In another aspect of the same subject "organics in drinking water," the use of carbon filters has been widely proposed as a means of reducing the level of organic precursors which permit the formation of chloroform following chlorination of drinking water supplies. The Safe Drinking Water Advisory Council had an opportunity to discuss this recommendation with Dr. Joshua Lederberg, noted scientist and Nobel Laureate of Stanford University. He cautioned against substituting unknown results for known reactions. He said that we do not know what the metabolism of charcoal or of elementary carbon is. He went on to say that the chemistry of what is in charcoal would make a horror story—carcinogenic polycylic hydrocarbons linked together by carbon bonds. In his opinion, it would take very small chemical modifications of charcoal *per se* to liberate such material. He wonders what happens when you expose charcoal to chlorine or to chlorinated compounds that are chemically active. He was not aware of any studies that are specifically directed to this particular concern.

In addition to the question Dr. Lederberg raises with respect to carbon filters, this comment concerns me because of our propensity to seek ready-made solutions to difficult problems without adequate background information or studies to guide us. The same caution must be addressed to our new-found interest in land disposal of wastewater and sludges. My reading tells me that there is much more unknown than is known about the significance of the quality of waste discharged to our underground aquifers and the ultimate fate of those wastes once they have been placed in the ground. Some studies have been conducted on the travel of some pollutants under some conditions. Few studies have fully addressed, and none have resolved, our current concerns for water reuse, groundwater recharge, discharges of highly concentrated effluents and sludges, or pretreatment requirements as related to the inherent relationship of these practices to long-term exposure of people to the trace elements ultimately associated with these wastes. In my opinion, we should approach land treatment of wastes with the utmost caution.

As for the relationship of wastewater effluents to water supply influents, we recognize that downstream communities have been treating and drinking discharges from upstream communities for a long while. What has changed is our knowledge of the character of the upstream discharges. Because of advances in laboratory technology, we are now able to discover and measure substances in our water which we did not know were there before. Even the most advanced wastewater treatment (AWT) technology, which in many instances can produce an effluent that is superior in every respect to the quality of water in some receiving streams, is not considered sufficient to permit the location of an AWT plant above a water supply intake.

The foregoing conclusion is possible when the ruling of the EPA on the construction of the Dickerson, Maryland AWT plant is analyzed. Along with other concerns, the public health considerations were reviewed, even though $12 million already had been spent in planning and designing the plant, and even though the site had been selected by EPA. After reviewing all available evidence on the issue, the Administrator's Executive Panel's final assessment was "though it would be preferable to discharge below the metropolitan area's water intakes, the health risks associated with the discharge from the proposed project do not, by themselves, warrant rejection of the proposed facility and denial of the grant application." Ultimately the grant application was denied. Now I am not arguing that the plant should or should not have been permitted. But if, as I believe, the concern on the part of EPA for the safety of the area's water supply figured prominently in the final decision not to support construction of the facility, I believe the public is entitled to the specific criteria that guided the government's conclusions relative to the protection of the public health. The basis for EPA's evaluation can be of enormous benefit to all of the communities that utilize the many surface sources around the country that receive treated sewage effluent and also serve as water supply sources.

I am aware that PL 92-500 requires the administrator, with respect to water quality-related effluent standards, to judge when the discharge from a point source or group of point sources "interferes with the attainment or maintenance of that water quality in a specific portion of the navigable water which shall assure protection of public water supplies." I am not aware of the published criteria or standards which enable him to make this evaluation.

It seems sometimes that industry is working at cross purposes with our efforts to clean up the environment and protect the health of people. Certainly, if asbestos, lead, kepone, PCB's and coke smelters are taken as examples where disposal of waste products is concerned, there is little to commend industry's willingness to voluntarily take the steps necessary to protect the public health. There is little in the history of our country or the economic system under which we exist to suggest that this is industry's role or that we should expect more from them in this regard.

Yet, there is a public feeling about corporate social responsibility and the relationship of industry to the community in which it exists, and the public on which it depends for its support and growth. In spite of this, and in spite of the lessons we should have learned from the use of DDT, mercury, cyclamates, lead, benzidine, diethylstilbestrol (DES), and many other toxic compounds, it required nearly six years to overcome industry pressure against enactment of the Toxic Substances Control Act. Had industry voluntarily recognized its responsibility in these situations and acted responsibly in the public interest, there probably would have been no need for this law. Under the circumstances, properly implemented, it can become the most significant piece of preventive health legislation enacted in this decade. However, from a fair and objective viewpoint, we must be concerned about the criteria that will be used to decide which toxic substances will be withheld from the marketplace to protect the public health.

In such extremes of thought, with respect to environmental control, that exist between what the public wants and what industry is willing to give, there must eventually be a meeting of the minds. Generally, this is best achieved when there is mutual understanding of what is required to meet desired objectives. I don't believe for one moment that industry sets out deliberately to harm the public health or that the public desires to require more of industry than is necessary for achieving protection of its health. I do believe that government can and should provide the leadership required to *establish the minimum practicable risk that produces the least acceptable harm to the public from environmental insults.* I do not believe government has done all that it can, nor all that it should, to reach this objective. The use of the term "government" in this instance is intended to include all levels of federal, state and local governments, the judicial, legislative and administrative arms of those governments, and not any one specific agency or unit.

Through these examples I have tried to stress the dilemma in which we all find ourselves in the environmental jungle associated with waste disposal. On the one hand, we have well-intentioned legislation that mandates action to protect the public health. On the other hand, we have very limited tools with which to accomplish this objective and even fewer tools with which to evaluate the effect of our actions. What we do to resolve this dilemma can have most important consequences on the health of our people and our future as a nation.

Supported by the foregoing, I believe that closer analysis may reveal that America's biggest problem—and indeed maybe its biggest enemy—is its ever-growing and yet unmanaged production and disposal of waste products. This apprehension is true whether our concerns embrace social, political or economic progress; ending of inflation and unemployment; increasing international trade and producing a balance of payments; expanding agricultural production or producing the goods desired for modern living; or creating the energy independence so important to our survival as a nation. The production of unwanted waste products is part and parcel to every action and reaction that accompanies proposed, accepted and implemented solutions to all of the societal problems that befall our communities and our nation. Our approach to these solutions affects us as a people and affects our place in the world in which we live.

A discussion of the tremendous volumes of waste produced in this country tends to boggle the mind. United States industries treat about 5,000 billion gallons per year of wastewater before discharging it to the environment. Of this volume, about 1,700 billion gallons are pumped to oxidation ponds or lagoons for treatment or as a step in the treatment process. Another 5,375 billion gallons per year of municipal wastewater is discharged to the environment, with an estimated 750 billion gallons of this amount discharged to the land. In the United States, municipal sludge production amounts to 4 million dry tons per year. Industrial sludge production is believed to be many times this amount. Some estimates place annual municipal solid waste production at more than 170 million tons per year and growing. To these figures must be added the millions of tons of gaseous waste that is produced annually; the untold hundreds of million tons of mining tailings disposed of each year; and the tremendous volumes (more than 24 million barrels per day) of oil field brines produced each day.

As a result of all the actions and interactions that accompany and follow the production of these wastes, we continue to create the kind of environment that fouls our air, pollutes our streams, desecrates our lands, and endangers our people. In the process, we create enmity between persons, problems between communities, disputes between states, and divisions between countries. We set people against industry, industry against government, and place the government, seemingly, against both industry and the people. These reactions are manifest in part by the many court suits associated with efforts to regulate waste disposal practices and to locate waste disposal sites at the local, regional, state and national levels. No one wants another's waste, and neither do we want our own.

In our zeal to correct the problems we have created for ourselves, we have almost legislated the nation into a no-win situation. We have a national air pollution control law that says you can't put the waste in the air, a water pollution control law that says that the nation's goal is zero discharge, and a safe drinking water law that says you cannot contaminate the domestic water supply. The only relatively unregulated medium available for disposal of our unwanted waste products is the land.

There are those who argue, "Why allow the discharge of any waste? Why not create the conditions under which all industry, public and private, recycles or reuses all waste products?" You and I know that the simple answer is that sooner or later there is an irreducible residual which, even under theoretical conditions, must be disposed of or stored in a safe repository. In actual practice, geography, economics, waste characteristics, and technology preclude total recycling and reuse of many waste products. In attempting to meet our disposal needs, the evidence suggests that once again we are blindly engaged in activities that could despoil two important and necessary resources—our land and the groundwater aquifers that exist underneath it. In our eagerness to make progress in dealing with the problems of the environment, and we are making progress, we seem to be pursuing a course, where land and waste products are concerned, of "out of sight, out of mind." We are permitting—and even promoting in some circumstances—under the guise of environmental conservation, or as accepted waste disposal practice, or under the practice of zero discharge of liquid wastes, the deposition of treated and untreated sewage and other hazardous wastes onto and into our lands and over our groundwater aquifers with little or, in some instances, no restraints.

Many years ago, a decision was made to use our streams, our rivers, our oceans and the air over our land for the disposal of the waste products of our progress. That decision was made in ignorance and supported by convenience. Waste volumes were small and disposal costs were low. Today we are paying the price for our lack of knowledge and absence of forethought. Notwithstanding the untold illnesses and deaths that some relate to these past activities, the many billions of dollars allocated from tax dollars and by industries to reclaim our streams and our air is testimony to the folly of our past actions. We must not allow these mistakes to be repeated in the future.

Early on, I made the statement that, in addressing our requirements for cleaning up our environment, we had almost legislated ourselves into a no-win situation—don't discharge wastes to the air, don't discharge wastes to the water. Now it appears that I am suggesting a zero discharge to the land. Lil Abner, Al Capp's famous comic strip character, probably would say, "As any fool can plainly see, it's got to go some place." And it must. Under the circumstances, we people have some hard choices to make. These choices concern beneficial use, risk analysis, and priorities associated with our knowledge of and appreciation for the interrelationship of air, land and water in our total environment. We must determine under what circumstances it is better for us to deposit our waste products on the land, in the air, or in the water. We must recognize that in the disposal of wastes convenience, expediency and low costs may be short cuts to environmental deterioration and endangerment of human lives. In the final analysis, we must accept the fact that knowledge and understanding of the forces of the environment with respect to the results of the sum total of our actions is essential to the protection of our people and the preservation of the earth.

Our efforts to date for accomplishing these objectives have primarily relied upon the application of best practicable and best available technology. While this approach is acceptable as a starter, it should not and must not be relied upon indefinitely. Such remedies on the face, and particularly in view of our nonknowledge with respect to health effects, must be compared to Russian roulette. They may be insufficient for the purpose, unnecessary at all, or overkill at best, to say nothing of the need for cost-effectiveness.

The accomplishment of the objectives I have cited will require, with respect to the environment, a rededication of purpose which is not now widely recognized. It will require an expanded and well-organized

environmental health effects research effort which, to my knowledge, is not underway or contemplated. It will require a time sense of urgency which seems to be waning as other crises push to the foreground. And it will require a substantial increased commitment of dollars and resources for environmental programs which to date has not been forthcoming.
I hope for the health and welfare of all of us, we rise to the challenge and find ourselves equal to the task.

3

THE SOCIAL, ECONOMIC AND POLITICAL IMPACTS OF ONLOT SEWAGE DISPOSAL

The Honorable Maurice K. Goddard

Secretary of Environmental Resources
Commonwealth of Pennsylvania
Harrisburg, Pennsylvania 17120

In 1966, the Pennsylvania General Assembly adopted the Pennsylvania Sewage Facilities Act, a rather comprehensive law dealing with a total program of sewage disposal. This legislation was passed in response to public health concerns over malfunctioning onsite sewage disposal systems; population growth; population trends showing movement away from the urban centers; and physical limitations on growth, such as poor soils and/or geology.

By adopting such legislation, the General Assembly brought together two primary facets of the program—planning and control. Planning provided the means of looking at the sewage disposal problem on a community-wide or even regional basis to meet future needs. It included not only a program for the installation of sewerage services, but also planning for utilization of onsite sewage disposal systems. Control was provided through a system of permits and inspections designed to insure satisfactory and safe installation of onsite systems.

It is important to note that the initial thrust of this legislation was preventive in nature and was concerned primarily with public health. Little emphasis was given then to the so-called secondary impacts of such a program. In fact, it wasn't until the early 1970s and the increasing environmental awareness of that time that secondary considerations such as social, economic and political impacts gained recognition in programs such as those established by our Sewage Facilities Act.

Advocates of limited growth or even no-growth began to realize the potential of the controls in the Sewage Facilities Act for exclusionary zoning. And development interests and individual property owners began to feel the threat of not being able to do what they wished with their property. It hasn't been until just recently that regulatory agencies have started to take a serious look at the social, economic or political impacts of new or revised programs.

Such impacts normally receive considerable attention in the planning and design stages for sewage collection and treatment systems. Federal requirements such as Section 208 and 201 Planning in the Federal Water Pollution Control Act Amendments mandate that such impacts be fully addressed. We must recognize, however, that such impacts also play an extremely important role in any program regulating utilization of individual onsite wastewater systems. Social, political and economic factors become even more important when one considers the increasing dependency on onsite methods of sewage disposal, often on a long-term basis.

In many respects, the public health concerns of the 1960s have given way to the 1970s' social and economic concerns which ultimately result in political impacts on an established regulatory program. It is difficult to separate completely the social and economic impacts of a given regulatory program. The political impact is somewhat easier to distinguish in that it usually is the result of social and economic considerations. Basically, however, the major social impact of regulating onsite sewage disposal systems is the alteration of individual rights. The major economic impact is the effect on the cost of land of such a regulatory program. And the resulting political reaction depends on how drastically those rights and costs have been affected.

In many respects, we are dealing here with individual rights versus group rights—the inherent constitutional right of free personal choice for the individual against the group rights relative to public health, environmental quality and aesthetic values. As one moves away from the cities into the semirural or rural areas, individual rights become more pronounced. However, the urban or semiurban population often looks upon the rural area and its amenities as worthy of preservation. And preservation of the rural integrity of an area directly affects individual rights.

The social and economic impacts of an onsite sewage disposal program often are manifested in land use issues. Land use is a broad concern that, in the past, has been overlooked by a variety of regulatory programs. But it no longer can be overlooked when discussing sewage disposal. Land use issues now are a "sign of the times" and frequently overshadow environmental health considerations which were paramount previously.

Let us consider a number of the social, economic and political impacts of onsite sewage disposal:

- Only a limited amount of land is suitable for installation of onsite sewage disposal systems. It has been reported that on a national basis only approximately 32% of the land area is suitable for onsite sewage disposal. Thus, onsite regulations which are uniformly enforced can place a premium on land and drive up land values in rural or semi-rural areas. Land values can increase to the point that certain income groups are priced out of the market. As a result, people begin to react, stimulating political involvement for compromise or change.

- When placing a premium on land by virtue of its suitability for development using onsite sewage disposal, land which is *not* suitable is devalued. Frequently this land, which is regulated out of the market, had been held for development purposes. Also, uninformed consumers often purchase land which is unsuitable for onsite sewage disposal. Once the developer or lot owner realizes his predicament, the timing of the forthcoming political response can be estimated almost to the hour.

- Strict enforcement of standards for onsite sewage disposal systems in areas of limited soil suitability can promote the installation of sewer services. This is dependent on pressure for growth in a given area. Sewers can induce additional growth in the community and, with time, alter the social structure of the community.

- Preservation of prime agricultural land has been the subject of considerable concern recently. Soils classified as prime for agricultural purposes frequently are the best suited soils for onsite sewage disposal systems; thus, growth can be induced in these areas. Such development can disrupt the agro-social community, not only with the primary impact of land use conversion but also with a multitude of secondary impacts such as highways, schools, commercial centers and a need for additional governmental services.

- Revision or expansion of onsite sewage disposal regulations to include alternate onsite methods can affect existing county or local comprehensive planning. The primary effect of an alternate system is that more land becomes suitable for onsite systems. Agencies developing comprehensive plans give full consideration to the physical characteristics of the planning area. Features such as soils, slopes and floodplains shape a municipality's plan for open space, recreation, highway corridors, residential development and other items. With the coming of alternate onsite systems, pressure is placed on existing comprehensive planning for change, often of drastic proportions.

- A very important political impact of any program (which is certainly inclusive of onsite sewage disposal) is the information gap between the technical complexities of regulatory agencies and the general public. To the lay person, government is extremely sophisticated and speaks in an unfamiliar language. The public has become frustrated, weary and impatient with bureaucracy, which it sees as evil, and is attempting to fight back through the political process.

- The economic value of limiting vector-borne disease or waterborne illness has a major impact upon the nation. The increased cost of medical care and public health administration can have a major cost reduction in communities which enforce an adequate program of onsite sewage disposal regulations. Lost work time, hospital costs, insurance premiums and other direct and indirect costs can be seen over the long term.

To give you an example of the political, social and economic factors which result from a strong onsite sewage regulatory program, I would like to discuss now my department's experiences with one specific community you all are familiar with—Gettysburg, Pennsylvania.

Gettysburg is primarily an isolated rural community of 8,000 people serviced by a severely limited and antiquated sewerage system. The community had escaped all forms of suburban expansion and remained virtually unchanged during the last 50 years.

Because of the lack of growth and the town's distance from major metropolitan areas, there never had been any municipal concern regarding upgrading and expansion of the sewage treatment facilities. The scattered residential development that occurred throughout the surrounding county did not justify any major concern.

With the trend of second-home ownership, however, either for recreational or retirement use, the Greater Gettysburg area was swept into the mainstream of development in the early 1960s. Construction of interstate highways established easy access to the area from Washington, D.C. and Baltimore. The aesthetically pleasing rural nature of the community and the low cost of land made the area an overnight Mecca for land development.

Immediately, the names of large recreational developments such as Lake Meade, Lake Heritage and Charnita became household words for the Gettysburg community. Overnight, the sleepy tourist town that hosted approximately four million visitors in the summer saw the financial gain which could be realized through this type of development. The social makeup of the community that had been mainly agricultural and tourist in nature was being transferred into the 20th Century.

Many residents saw their community as encroached upon by national franchise, building-related business and all the sophistication of downtown Washington. Others saw an opportunity to buy into land developments and businesses that would prosper by this turn of events. A social-economic dichotomy had developed between those who felt their community should be static and those who wanted the benefits of growth.

It was at this point that the onsite sewage disposal program crossed the existing and planned socioeconomic community. With enactment of the Sewage Facilities Act, parameters were established for onsite sewage disposal systems. These regulations greatly limited development in Gettysburg and surrounding areas because of severe soil limitations. Unfortunately, many thousands of lots were marketed in the area prior to enactment of the legislation. And many prospective home builders held land which was worthless for its intended purpose because of its unsuitability for sewage disposal.

Also, many financial institutions were adversely affected, either through investment or mortgage loans. Many prospective businesses either lost money or failed to materialize. The financial gain which was anticipated actually was lost.

The obvious solution to this dilemma seemed to be expansion and extension of sewerage services. But rural attitudes and community rivalries slowed any attempts to construct and expand conventional wastewater collection and treatment systems. The entire community hung suspended, unable to move forward because of the onsite sewage regulations and the unwillingness of communities to collectively plan for sewers.

Finally, a major effort was undertaken to alleviate the problem through construction of a regional treatment facility for the Greater Gettysburg area. After 10 years of difficult negotiations, municipalities agreed to make the commitment. Contracts were let to engineering consultants and final preparations are being made to sewer the area.

Once again, as one might expect, development pressures are running rampant. New subdivisions and businesses are being planned in anticipation of the financial reward to be derived through installation of sewers. The concern over land use for this national shrine once again is in a position to be sacrificed through further inducement of growth.

The saga of Gettysburg is not unlike the story of other communities across the nation. This example, like those stories, points out the problems of social, economic and political impacts of change brought about by regulatory programs. In summary, regulatory agencies, whether at the state, regional or county level, must be aware of social, economic and political impacts of any new or revised programs and react accordingly.

This is particularly true if meaningful programs are to be developed and implemented for control of onsite wastewater systems. Agencies must deal effectively with the realities of the socioeconomic and political processes. These realities can be influenced by increased public awareness of the need for and implementation of various programs. An essential part of any successful program dealing with onsite wastewater systems is a commensurate public information effort.

Any agency becoming involved in the onsite sewage disposal field must take into account a variety of federal regulations, programs and policies. The Federal Water Pollution Control Act is one such mandate—through Section 208 and 201 Planning—that is certain to intensify the role of onsite controls and related land use considerations.

The controversy over individual versus group rights is certain to intensify in coming years, and it is likely that environmental groups will be at the heart of this debate. Amidst all the controversy and related impacts, we must realize, as recently stated in *The Public Interest*, that "The elite's environmental deterioration often is the common man's improved standard of living."

4

U.S. EPA RESPONSE TO PL 92-500 RELATING TO RURAL WASTEWATER PROBLEMS

Office of Research and Development (Sections 104 and 105)

James F. Kreissl

    Sanitary Engineer
    Municipal Environmental Research Laboratory
    Wastewater Research Division
    U.S. Environmental Protection Agency
    Cincinnati, Ohio 45268

HISTORY

Section 104(q)(l) of PL 92-500 directs the U.S. EPA to conduct a comprehensive program of research and development into new and improved methods of preventing, reducing, storing, collecting, treating or otherwise eliminating pollution from sewage in rural and other areas where conventional sewage collection systems are impractical, uneconomical, or otherwise infeasible, or where soil conditions or other factors preclude the use of septic tank and drainage field systems.

At the time of the subject legislation, the primary effort of the then "Non-Sewered Wastes Program" within EPA's Office of Research and Development was related to the demonstration of pressure sewer systems. One other task had been completed which involved cataloguing several different onsite wastewater systems and commercially available water reduction devices for home application, along with an attitudinal survey toward such devices. These projects were clearly related to the intent of Section 104(q)(l).

In fiscal year (FY) 1973 a study of septic tank pumpings (septage) characterization and treatment was initiated at the EPA pilot plant facility in Lebanon, Ohio. Also, ongoing pressure sewer demonstration studies continued to be monitored during this fiscal year.

Fiscal year 1974 funding of the present "Small Flows Research Program," which is an outgrowth of the "Non-Sewered Wastes Program," represented a marked increase over that of previous years. This increased level of funding permitted embarkation into two major projects. The first was designed to accomplish two purposes: (1) the demonstration of a different type of pressure sewer system, and (2) the demonstration of a vacuum system in a single location. The second project was designed to meet the program need to investigate thoroughly onsite treatment and disposal systems. The most comprehensive study anywhere in the country at that time was underway at the University of Wisconsin, and was in critical need of new funding. The marriage of needs was consummated with resultant funding of the first year of a three-year program. Meanwhile, the septage study at Lebanon was continued, while early pressure sewer demonstration studies were phasing out.

In FY 1975 the higher level of funding initiated in the previous fiscal year was continued. Because of this, not only were the onsite studies at the University of Wisconsin continued at a consistent level of funding, but a modest second project on evapotranspiration and mechanical evaporation systems was initiated at the University of Colorado. A new initiative in the area of improved methods of wastewater collection was undertaken with the funding of a study of vacuum sewer systems in the United States. The objective of this study was to critically analyze design methods, construction considerations, O/M requirements, costs, and limits of applicability. A study of septage handling techniques in selected areas of the country was initiated, along with a small pilot-scale study of septage addition to activated sludge plants at the EPA pilot plant in Blue Plains, D.C.

The FY 1976 funding level represented about a 30% reduction from the previous two years. However, the third and final year of the Wisconsin study was funded and a comprehensive large-scale pilot and full-scale septage treatment and disposal study was initiated. A small amount of supplementary funding was also added to the vacuum/pressure demonstration project to help defray inflationary costs.

PRESENT AND FUTURE STATUS

At the onset of FY 1977, the EPA has clearly indicated a renewed interest in the Small Flows Research Program as evidenced by a substantial increase in proposed funding for not only this year, but future years. The concept of conventional sewers for smaller and sparsely developed communities is now being seriously challenged by both federal grant-approving authorities and by local communities who must pay exorbitant user costs for the "privilege" of being sewered. Because the cost of sewering by conventional means can vary from two to four times (on a per household basis) that found in larger or more densely populated communities, the need for feasible alternatives is clear.

Due to the comprehensive scope of the Small Flows Research Program and the recognized need for reliable design, and capital and O/M cost information, this program will be directed toward solving major problems and providing as much useful information as possible over the next few years to impact contemporary planning, engineering and implementing practices for individual home and small community systems. As a part of this effort some new initiatives will be made in the areas of onsite system alternatives in order to properly assess newer concepts of wastewater manipulation, reuse and conservation in technological, economic and social terms. Some projects will be undertaken to complete the R&D effort for other initiatives which are well along toward the goals originally set for them. These areas include the new sewering techniques, alternative onsite systems with traditional household wastewater generation patterns, and septage handling. Some long overdue treatment studies for small communities will also be initiated.

The status and direction of the Small Flows Program is best discussed in the context of its subelements. Each area of activity is in a different stage of progress at this time, in terms of the goals set for that activity.

Advanced Collection Technology

The results of the new collection systems studies have for the most part satisfied the goals set for this activity. The overall goal is to produce reasonable guidance for planners, engineers, municipalities (or districts) and regulatory agencies on the use of pressure and vacuum sewers as a feasible alternative to conventional methods. Previous EPA-ORD projects at Albany, New York; Phoenixville, Pennsylvania; and Grandview Lake, Indiana, along with several non-EPA installations have generated considerable data on the design, capital costs, waste characterization, and short-term operation and maintenance (O/M) costs of pressure sewers. More remains to be learned about the treatment of the wastes generated, long-term O/M requirements of these systems, and the ultimate applicability of this technology. A project is scheduled for FY 1977 to answer these outstanding questions.

Based on the information gained by manufacturers of pressure sewer equipment, EPA project reports, and designers and operators of some early systems, several pressure sewer systems are being installed around the country at this time. The adoption of this technology has been limited to relatively few competent engineering firms who have tackled jobs which defied traditional solutions, *i.e.,* conventional sewers and septic tank-soil absorption systems. Unfortunately, the number of engineering firms that conscientiously evaluate this technology as an alternative to traditional solutions are few in number.

Previous pressure sewer projects have shown the cost-effectiveness potential of this technology. For example, at Grandview Lake, Indiana, the capital cost estimate for conventional sewerage exceeded $10,000 per existing home, while the actual cost of pressure sewerage was less than $2,500 per home. In Douglas County, Oregon, estimates of capital costs for a conventional versus a pressure sewer showed a 50% savings for the pressure system. One existing pressure sewer system in Priest Lake, Idaho, was built for less than 10% of the estimated cost of a conventional system.

The actual cost of any sewer system is a function of many factors including the population density of the region. Needless to say, no centralized collection system is feasible for rural areas where the distance between homes is excessive. However, the use of this technology for recreational communities, urbanizing or suburban areas, and low-lying or difficult to serve districts of larger communities should be given proper consideration by any design engineer interested in providing a cost-effective solution to his client. The pressure sewer technology is particularly advantageous in areas where rocky conditions cause tremendous increases in sewer costs for each additional foot of excavation depth; where high groundwater poses serious excavation problems which require pumping and shoring in deeper trenches; where hilly terrain would require numerous lift stations and force mains or excessive depths of cuts for conventional sewers; and where the population density is low and excessive lengths of sewer between homes are required.

Similar arguments have been made for the adoption of vacuum sewer technology. Like pressure systems, vacuum sewers require minimal depths of excavation and employ small-diameter plastic sewer pipe. The major differences between the two systems are the need for pockets in the vacuum mainline which dictates closer attention to grade during construction and the motive force application. Pressure systems require a pressurization facility, pump or grinder-pump (GP) at each inlet, while the vacuum system employs a centralized source of vacuum on the collection line with a valve at each inlet that operates off the vacuum in the main collection line.

The economics of vacuum systems are not as readily available as those of pressure systems, primarily due to their use by developers rather than by municipalities. However, at least two instances of municipal vacuum sewers are known in locations where flat terrain and high groundwater problems exist (Mathews, Virginia; and Plainville, Indiana). In both cases, the systems were installed because of lower capital costs. In the case of Mathews, Virginia, the capital cost of the vacuum system was $376,000 compared to a $644,000 estimate for a conventional system.

Presently, small vacuum and pressure systems have been designed and constructed in the City of Bend, Oregon, as part of an EPA R&D grant. These systems are being monitored to determine their performance, O/M requirements and user acceptability. Capital cost information has been determined for these small ($\sim$ 12 homes on each) systems, but it will require interpretation for general applicability to realistic-sized installations. Also, a contract to evaluate the state-of-the-art of vacuum sewer systems is nearing completion, and a report will be available in 1977.

Onsite Alternative Systems

The development of onsite systems as alternatives to conventional septic tank-soil absorption systems (ST-SAS) is proceeding in several locations. New system development and testing is proceeding in parallel with studies that increase our knowledge of the functions of conventional systems, with an overriding concern for the legal and managerial concepts required to optimize the performance of these alternatives.

Conventional ST-SAS are not presently well understood, as witnessed by the tremendous variability in state codes which govern their design and application. A study at the University of Wisconsin, with U.S. EPA and state funding, has greatly increased our understanding of how these systems operate and why they may fail, and has provided significant data on some alternative approaches for onsite treatment and disposal of wastewaters.

To understand the onsite wastewater problem it is necessary to consider the three phases of the overall problem: (1) wastewater generation, (2) treatment, and (3) disposal.

The first step, *i.e.*, a study of wastewater generation and composition, has been accomplished.[1] The results indicate that approximately one-third of the total wastewater volume is generated by the toilet. This fraction is called the "blackwater," and it contains about one-half of the organics and solids, three-fourths of the nitrogen, one-fifth of the phosphorus, and most of the fecal bacteria and virus contained in the total wastewater. The remaining two-thirds of the wastewater (greywater) contains the balance of the above pollutants. Equally important is the household hydrograph, or hourly variation in the wastewater flow and constituent matter. A typical hydrograph is shown in Figure 1.[2]

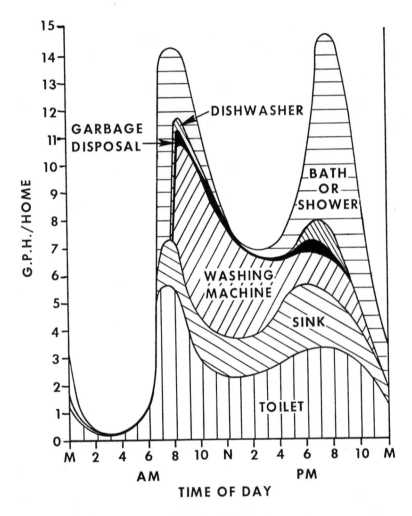

Figure 1. Daily household water use.

If one assumes normal wastewater generation and composition and treatment by an adequately sized and designed septic tank, several alternative methods of disposal are available. If disposal alternatives are limited to (1) conventional subsoil, (2) improved subsoil, (3) surface discharge, (4) evapotranspiration, and (5) mounds, the first choice is the conventional ST-SAS. Although millions of these systems are working throughout the country, there are several problems inherent to these systems which restrict their applicability to areas that are generally devoid of one or more of the following conditions: (1) thin soil over creviced bedrocks; (2) highly impermeable soils; (3) high groundwater conditions; and (4) thin soil over impermeable strata on steep slopes.

The University of Wisconsin study has pointed out several major problems that contribute to early failure (<3 years) of ST-SAS. These are: (1) poor site evaluation; (2) inadequate review by regulatory agency; (3) poor construction; (4) inadequate inspection; and (5) hydraulic overloading after construction. If this early failure can be avoided, ST-SAS systems may be expected to have substantial (>20 years) service lives.

Site evaluation and construction techniques have been investigated at the University of Wisconsin to show the pitfalls of present evaluation techniques, and new approaches have been evaluated and described. However, the value of any technique is only as good as the regulatory agency which controls installation of onsite systems. This institutional aspect of the problem has also been investigated, and improvements have been suggested which would generally improve conventional ST-SAS performance and have equal or greater applicability to new alternative systems.

The functioning of conventional ST-SAS as it relates to chemical and hydraulic phenomena and to microorganism removal has been studied extensively at the University of Wisconsin. Failure, defined as inadequate hydraulic acceptance or purification of wastewater, has also been characterized. The basic elements of concern are, therefore, becoming more clearly defined as a result of this work.

The better understanding imparted by these studies indicates that the soil absorption system should be designed as a function of the soil. Although present practice does attempt to change the size of these fields, the basic design is unchanged between different soils. Improved designs employ such things as pressurized distribution to maximize soil purification and dosing to maximize soil field life cycles. Although more work remains, these systems are generally ready for application.

The problems of existing system failures are also being addressed. Several successful rejuvenations of failed soil systems have been reported by the University of Wisconsin researchers by judicious application of hydrogen peroxide. Although much more development work is necessary, the potential value of this technique is substantial when compared to the time and dollar requirements of the present approach of constructing a new soil absorption system (SAS).

Alternating beds have been known to permit long-term system life. Unfortunately, no scientific study has ever been undertaken to determine how these beds should be sized or how often they should be alternated to insure complete recovery of their capacity to receive septic tank effluent. Such a study will be initiated in 1977.

Surface discharge of septic tank effluent would be completely unacceptable from public health and aesthetic points of view. However, studies at Wisconsin have shown that intermittent slow sand filtration at realistic rates can consistently produce an effluent meeting a 10 mg/l BOD, 10 mg/l SS standard. Disinfection of that effluent with chlorine or UV light has been successfully demonstrated. There are three drawbacks to this type of system. First is the increased O/M requirement, which could only be handled by a centralized management arrangement. Second is the potential monitoring problem represented by a proliferation of these systems in an area. Last, but not least, is the cost of such a system.

Evapotranspiration (ET) as an alternative disposal mechanism is feasible in some western areas of the country. Presently, an EPA-sponsored study at the University of Colorado is attempting to optimize the design of ET beds in their classic form and in the so-called "ETA" form where soil percolation is encouraged to assist the ET phenomena. An additional aspect of this study is designed to study prototype mechanical evaporation devices which may permit a much wider application of evaporative disposal systems in areas where excessive rainfall presently precludes the use of sealed ET beds.

Mound systems have been extensively studied at the University of Wisconsin and elsewhere. Their primary utility is their ability to (1) provide purification of effluents in areas where creviced bedrock is covered with only a thin soil mantle, and (2) effectively remove the soil system from high groundwater conditions in areas where this condition predominates. The main drawbacks to mounds are their cost and negative aesthetic value.

Many of the alternative onsite systems require further study to verify their functioning, optimum design, capital costs, and O/M requirements. Additional evaluation of this type is planned during the next few years. However, the available information is generally sufficient for immediate successful application of technologies to solve specific local problems.

Another consideration for onsite systems treating normal household wastewaters which is worthy of note is the use of aerobic devices instead of septic tanks for pretreatment of wastewater prior to disposal. The University of Wisconsin study has shown that although the aerobic devices are generally capable of producing a better quality effluent than septic tanks, the beneficial effects of that effluent on reducing clogging of marginal or problem soils is not significant. This is also true for evapotranspiration systems because the evaporative process occurs at essentially the same rate for both effluents. The area where the aerobic systems may represent a cost-effective approach is for surface discharge. If these systems are capable of producing an effluent which either meets or requires only minimal polishing to meet discharge standards, this could represent a cost-effective surface discharge system. Past studies have not been encouraging due primarily to a lack of necessary O/M required by these systems. A properly organized management program could overcome this problem, if overall economics are favorable.

The onsite system which is least understood at this time is that which begins with wastewater manipulation within the home. Systems proposed generally employ either complete elimination or drastic reduction of blackwater, or some form of recycle of greywater. The primary question to be answered when wastewater manipulation is contemplated is the question of public acceptance. Public attitude toward these systems is difficult to judge and is subject to innumerable influences, *e.g.,* water availability, nature of local environment, etc. A survey performed in 1969 by the General Dynamics Corporation, Electric Boat Division, questioned more than 350 homeowners in 9 locations in the United States about water savings fixtures.[3] Over 75% replied that they did not object to low-flush and dual-cycle toilets, but the majority rejected home urinals and recycle toilets, such as those used on commercial airliners. Intuitively, one would suspect that drastic water supply shortages or overwhelming wastewater disposal problems would alter these responses, but the degree of alteration or limits of public acceptability cannot be quantified without knowing local conditions.

The most popular wastewater manipulation scheme being proposed is the use of waterless toilets. The technical validity of this approach lies in the answer to the following questions:

1. Does the toilet function in an acceptable manner without endangering public health or representing an aesthetic nuisance?
2. How is greywater treated and eliminated?
3. What are the installed costs and O/M requirements?

Presently, there is a need for data to satisfy the first question. Although several toilet designs are currently employed in recreational homes, cabins and cottages in other parts of the world, there is insufficient information on most systems to make any judgment on their applicability for more concentrated use in the United States. A 1977 study is scheduled to develop information on these systems.

Greywater has been shown to require treatment and disposal in much the same manner as the total wastewater flow. However, quantification of its treatability, soil clogging potential, etc., needs more study. Some data will be forthcoming from the University of Wisconsin study, but additional information will be required as part of the project described immediately above.

Recycle systems that are presently available generally utilize the principle of recycling wash waters (greywater without the kitchen sink contribution) to the toilet or to an outside faucet for lawn sprinkling. These systems will also be studied as part of the wastewater manipulation program in the coming year.

Septic Tank Sludge Handling

The millions of septic tanks scattered around the country accumulate solids in the forms of sludge and scum which must be removed periodically to protect soil absorption systems. These accumulations and the liquid content of the septic tank are called "septage," which is produced in quantities estimated at 4 billion gallons per year in the United States.

Studies from EPA pilot plants and grant projects have produced the bulk of available septage characterization data. Three methods of septage handling have been studied thus far, and several other methods are to be studied in a new grant project started in the fall of 1976. A nationwide study of septage handling has been completed and will be available in early 1977. A full-scale monitoring project will also be initiated in 1977.

Characterization data from the EPA pilot plant studies are shown in Table I.[4,5] Heavy metal content of septage is compared to those of certain municipal sludges in Table II.[6] Table II supports the widespread use of land application methods for septage disposal, where land is available for this purpose. Regional septage facilities and sewage treatment plant disposal methods will be employed in most areas where land disposal is not feasible, *e.g.,* insufficient land availability, high groundwater contamination potential, etc. All of the ongoing and future studies will concentrate on developing proper design criteria for septage handling facilities.

Miscellaneous Studies

The Small Flows Research Program is also concerned with small community treatment systems, recreational waste treatment, and non-sewered commercial wastewater treatment. Although some work has been completed in each of these categories of concern, lower priorities have been placed on these areas due to a lack of resources within the program. However, studies are planned to begin in these areas within the next two years to provide necessary information for upgrading present practices in each.

Table I. Septage Characterization

Constituent	Concentration (mg/l)	
	Arithmetic Mean	Range
TS	38,800	3,600-106,000
TVS (%)	65	32-81
TSS	13,000	1,800-22,600
VSS (%)	67	51-85
BOD_5	5,000	1,460->18,600
COD	42,850	2,200-190,000
TOC	9,930	1,300-18,400
TKN	677	66-1,560
NH_3-N	157	6-385
TP	253	24-460
pH (units)	—	6.0-8.8
Grease	9,000	600-23,500
LAS	157	110-200
Fe	205	3-750
Zn	49.0	4.5-153
Mn	5.0	0.5-32
Cd	0.71	<0.05-10.8
Ni	<0.9	0.2-3.7
Hg	<0.28	<0.0005-4.00
Se	0.08	<0.02-0.3
Cr	1.1	0.3-2.2
As	0.16	0.03-0.5
Cu	6.4	0.3-34
Al	48	2-200
Pb	8.4	1.5-31

Table II. Heavy Metal Content of Septage and Sludge (mg/kg)

Metal	Septage	Salotto	Other U.S.	Denmark	Sweden
Cd	5.5	43	69	10	9.8
Cr	21.0	1,050	840	110	170
Cu	28.1	1,270	960	340	670
Hg	<0.24	6.5	28	7.8	5.8
Mn	106	475	400	350	400
Ni	<28.5	530	240	37	65
Zn	1,280	2,900	2,600	2,600	1,900

Costs

Although the present information on some of the alternatives is limited at this time, an increasing number of these systems are being constructed in areas where conventional onsite systems have failed. The data bases generated by the full-scale alternative systems are still relatively sparse, but they do provide enough information to make some crude comparative cost estimates for several alternative wastewater systems.

Conventional septic tank-soil absorption systems sized in accordance with the *Manual of Septic Tank Practice*[7] generally cost from $800 to $2000 and require about $10 per year of O/M in the form of periodic pumping. If the tank and absorption field are estimated to have service lives of about 25 years, the annual cost of these systems at a municipal interest rate (assumed because of the construction grant eligibility of municipally owned and operated onsite systems) of 5.87% is between $80 and $150. The low value corresponds to permeable soils (10 minutes/inch percolation rate) while the high value is for marginally acceptable soils (60 minutes/inch). Higher and lower costs have been experienced, but the above range is representative and is based on excavation costs of about $1.25 per square foot, trench depths of less than 4 feet, and installed septic tank costs of approximately $350.

With similar assumptions and some scattered data for actual installations, some additional cost estimates for other alternative systems are shown in Table III. Many of the alternative system costs shown in the table are based on minimal data. Therefore, they are presented only as a rough guide to illustrate relative costs for determining an economic order of consideration.

Some tentative trends can be seen from the information in Table III. If no extreme problems exist in the area under consideration, *e.g.*, thin soil over creviced bedrock, highly impermeable soils, steep slopes or high groundwater, it would appear that conventional, dosing or alternating bed systems are clearly the most economical choices. If problems like high groundwater or impermeable strata at a shallow depth are prevalent, these systems are not feasible and other solutions such as mounds and ET beds become more attractive, despite their higher costs. In highly impermeable soil regions with steep slopes, the ET beds and surface discharge alternatives become the only feasible alternatives.

Two assumptions are made in the above trend identifications. These are:

1. Only new homes are considered.
2. Sewering is not a feasible alternative.

Table III. Onsite Alternatives Cost Estimates

Alternative	Soil	Capital Cost $	O/M Cost $/yr	Total $/yr
Septic tank–conventional SAS	Good	975	10	85
	Fair	1288	10	109
	Poor	1600	10	134
Septic tank–pressurized distribution (dosing)–SAS	Good	1317	35	134
	Fair	1641	35	159
	Poor	1964	35	194
Septic tank–alternating beds	Good	1700	10	141
	Fair	2326	10	189
	Poor	2950	10	238
Septic tank–mound	—	3500	35	305
Septic tank–(ET) evapotranspiration	—	4000	10	319
Septic tank–sand filter–disinfection	—	3415	150	421
Aerobic unit–pressurized distribution (dosing)	Good	2347	122	326
	Fair	2671	122	351
	Poor	2994	122	376
Aerobic unit–sand filter–disinfection	—	3395	207	492
Aerobic unit–disinfection	—	2645	147	374

If an existing home has failed conventional ST-SAS, some of the alternative actions available are:

1. Install a new soil system, if space is available on the property.
2. In addition to the above, add an alternating valve to the system to permit future alternation of beds.
3. Hydrogen peroxide treatment of the failed system.

The first action is the one most commonly used when sufficient space is available on the homeowner's property. Actually, the second alternative is far more sensible in that it essentially insures long-term performance by utilizing the proven concept of alternation at a cost increase of no more than 10% over the first option. Hydrogen peroxide treatment has not yet been fully studied, but it may represent the least costly alternative if the exact location of the existing trenches is known. The cost of a new SAS is dependent on the soil percolation rate in most areas, and generally varies from $500 to $1200. The addition of an alternating valve costs an additional $100, while an idealized hydrogen peroxide treatment should run about $300 to $400.

The wastewater flow reduction alternatives have not been well documented in terms of costs. Toilet equipment and delivery costs are available from manufacturers, but installation cost can only be estimated. Since the simplest and most allowable (from a regulatory standpoint) greywater disposal method is likely to be a ST-SAS, some rough cost estimates can be made for a total household system, assuming the units would be acceptable from a public health and aesthetic viewpoint and that proper maintenance will be provided by the homeowner. Based on the further assumptions that a reduced requirement exists for ST-SAS for greywater alone and that no credit is given for water savings because of the likelihood that private wells are the source of water supply, some cost estimates are shown in Table IV. All of the above assumptions are subject to argument, but some interesting trends do result from the table.

Table IV. Water Savings Alternatives Cost Estimates

Alternative	Soil	Capital Cost $	O/M Cost $/yr	Total $/yr
Large biological toilet + ST-SAS	Good	2900	20	244
	Fair	3200	20	268
	Poor	3500	20	291
Small biological toilet + ST-SAS	Good	1900	90	237
	Fair	2200	90	260
	Poor	2500	90	283
Incinerator toilet + ST-SAS	Good	1650	180	308
	Fair	1950	180	331
	Poor	2250	180	354
Low-flush toilet + ST-SAS	Good	1400	96	210
	Fair	1700	96	234
	Poor	2000	96	257

The low-flush toilet appears to have some cost advantage when compared to waterless toilet systems. Among the waterless approaches, both the large and small biological toilets show some advantage over the incinerator toilets due to their significantly lower operating costs. The costs of the flow reduction alternatives generally lie between the normal trench alternatives (conventional, dosing, pressurized distribution and alternating beds) and the more costly mounds, ET beds and surface discharge solutions.

From Tables III and IV it is again apparent that more conventional solutions are advantageous for most individual home applications from an economic standpoint. However, local conditions of water shortage, groundwater contamination, poor soil conditions, etc., may obviate those solutions and provide justification for the more costly alternatives. A study is planned for 1977 which will thoroughly investigate the applicability, sensitivity, and limitations of each alternative to provide guidelines for future cost-effective analyses for rural communities and to help pinpoint the research areas on which future projects should concentrate. Simultaneous work will be devoted to the proper assessment of several types of household devices for water conservation, recycle, etc., to determine their utility, reliability, and overall applicability to non-sewered dwellings.

If the contents of Tables III and IV are applied to rural communities, the alternative against which these systems must be judged is that of centralized collection and treatment. Collection may be accomplished by conventional, pressure or vacuum sewers. The annual cost per home of conventional sewerage as a function of population density in smaller communities as documented by a survey of actual costs conducted by the EPA Construction Grant Program,[8] can be expressed as:

$$\text{Annual Cost (\$/yr)} = 516 \, e^{-0.1 \, (\text{persons/acre})}$$

No comparable set of cost data presently exist for pressure and vacuum sewer systems. However, in rural communities with population densities of approximately five persons/acre or less, the alternative systems are often significantly advantageous from an economic standpoint. Given local conditions such as high groundwater, hilly topography and rocky terrain, this advantage can be greatly enhanced.

SUMMARY

The Small Flows Research Program of the U.S. EPA-ORD has made several strides toward impacting rural wastewater problems within the limits of its resources. As a consequence of the work already undertaken and that which is planned, information will be generated and disseminated which will permit engineers, state and local regulatory agencies, and rural communities to properly evaluate the most cost-effective solutions to their rural wastewater problems.

REFERENCES

1. Witt, M., R. Siegrist and W. C. Boyle. "Rural Household Wastewater Characterization," in: *Home Sewage Disposal*, ASAE Publication Proc. 175 (December 1975).
2. Bennett, E. R. and K. D. Linstedt. "Individual Home Wastewater Characterization and Treatment," Colorado State University Environmental Resources Center Completion Report Series No. 66 (July 1975).

3. Bailey, J. R., R. J. Benoit, J. L. Dodson, J. M. Robb and H. Wallman. "A Study of Flow Reduction and Treatment of Waste Water from Households," U.S. EPA Water Pollution Control Research Series No. 11050 FKE (December 1969).
4. Feige, W. A., E. T. Oppelt and J. F. Kreissl. "An Alternative Septage Treatment Method: Lime Stabilization/Sand Bed Dewatering," U.S. EPA Environmental Protection Technology Series Report No. 600/2-75-036 (September 1975).
5. U.S. EPA Internal Reports.
6. Salotto, B. V., E. Grossman and J. B. Farrell. "Elemental Analysis of Wastewater Sludges from 33 Wastewater Treatment Plants in the United States," in: *Pretreatment and Ultimate Disposal of Wastewater Solids*, U.S. EPA Publication No. 902/9-74-002 (September 1974).
7. U.S. Public Health Service. *Manual of Septic Tank Practice.* Publication No. 526 (1969).
8. Dearth, K. Private communication.

5

U.S. EPA RESPONSE TO PL 92-500 RELATING TO RURAL WASTEWATER PROBLEMS

Office of Water Program Operations (Section 201)

Keith Dearth

> Municipal Constructions Division
> U.S. Environmental Protection Agency
> Washington, D.C. 20460

Section 201 of the 1972 Amendments to the Water Pollution Control Act provides grants for planning and construction of municipal wastewater treatment facilities. Congress authorized $18 billion in grant funds for fiscal years 1973-75 and the EPA is planning for a $5 billion/year program in the future. This level of funding makes the grants program the largest public works program in the nation.

To administer Section 201 EPA has established a three-step grants procedure covering all phases of the wastewater treatment works construction program. These steps are facilities planning (Step 1), final design and preparation of plans and specifications (Step 2), and actual construction (Step 3). These grants provide 75% of the eligible costs involved and must be applied for prior to the commencement of each step. The funds may, under certain conditions, be spent for small wastewater treatment systems, servicing equipment and residual waste disposal facilities.

Initially, communities whose wastewater systems are sources of pollution prepare a plan of study and submit it with an application for a Step 1 facility planning grant to EPA through the state. This plan of study describes the local needs for treatment works, the scope of the required facility planning effort (including a list of principal tasks to be accomplished), a schedule and estimated costs. The facility plan is normally prepared by the community's engineering consultant. The state provides the effluent limitations to be met by the treatment. Starting with this limitation, the consultant further considers the existing problems, assesses the current and future situations, develops and evaluates alternatives and selects the cost-effective plan. During this step, evaluation of monetary costs and the environmental evaluation of alternatives as well as public participation are required. The plan must include assurances that the community can carry out the plan and bear the local costs.

A Step 2 grant application together with a completed facility plan is submitted to EPA. Following approval, a Step 2 grant is awarded for preparation of plans and specifications. After these documents have been prepared and approved by state and EPA officials, the Step 3 grant is awarded and the process of actual construction begins. Of course this procedure is governed by the state's priority list and is subject to availability of funds.

For purposes of this paper we are concerned primarily with the facility planning phase, Step 1. This is the phase where the engineer develops and evaluates alternatives and selects the cost-effective plan. Many of the facility plans reviewed by EPA Headquarters, to date clearly demonstrate that the no-action or minimal action alternative has not been carefully considered even where such an option might be practicable. In some cases, existing onsite and other small facilities have been scheduled for abandonment without sufficient justification, and costly collection and conveyance systems with traditional secondary plants or even advanced waste treatment plants have been prescribed. In some instances, the collection systems have not been included on the state priority list for grant funding. This has meant that the local community has had to finance 100% of the capital costs for collector sewers. The resultant high debt retirement costs are sometimes beyond the capacity of small communities to pay.

New treatment works operation and maintenance costs often by themselves are very high for median- and lower-income families in some rural and semirural communities.

One low-income community in a northeastern state slated abandonment of its onsite systems and constructed a collection and advanced treatment system. Though the state and EPA paid the entire cost of the plant, the operation and maintenance costs and the collection system debt retirement costs total approximately $200 per year per family. Over half the families now either refuse to hook into the system or refuse to pay their user charges. A nearby community with its new AWT plant is also paying $200 per year per family just for operation and maintenance costs alone.

To attempt to determine the scope of the problem EPA recently studied 258 facility plans for pending projects from virtually every state. The survey indicated that operation and maintenance plus debt retirement of the local share for recommended new facilities will cost in excess of $100 per household per year in 40% of the communities and $200 per household in 10% of the communities. The major problem arose when of these 258 plans, 83 called for completely new collection and treatment systems. Three-quarters of the 83 indicated costs in excess of $100 per year per household and one-fifth in excess of $200 per household per year. Costs will exceed $300 per household in several instances. Communities under 10,000 persons in general seem to have the most serious problem. This is partly because they cannot benefit from economics of scale in construction and operation of facilities. It is also because often new facilities must be built to serve the entire community, whereas larger communities often have some facilities in place.

How can this problem be solved? Consideration and selection of the truly cost-effective alternative will help provide an answer.

Adequate evaluation of the no-action or minimal action alternative for the non-sewered or partially sewered community must be made and presented in all facility plans as an option to collection system construction. Where onsite wastewater systems are allegedly causing pollution of ground or surface waters, evidence in the form of stream sampling and discharge data or documentation from appropriate health officials must be provided citing the specific conditions. Also, where such systems are malfunctioning the feasibility of correcting the problem must be carefully evaluated. Other pertinent questions to be explored are:

1. How many onsite systems are there and how many are malfunctioning?
2. What is the nature of the malfunctioning and where are these units located?
3. When was each system installed?
4. What were the results of the percolation test(s) before installation? What are they now?
5. How frequently has each unit been pumped out? How frequently inspected? Is there a garbage grinder connected into the system?
6. Describe the physical areas, including soils and groundwater characteristics, where malfunctions are occurring. What is the depth to the groundwater table?
7. Would a local commercial or municipal organization with adequate pumping vehicles and a treatment facility properly utilized eliminate most of the malfunctions of the onsite systems?
8. Would a municipal organization be able to obtain ownership of, or easements to, each onsite system?
9. What are the costs of upgrading the systems, of providing maintenance service, etc.?

10. In areas where conditions vary and only a portion of the area is suitable for septic tank service, will a limited collection and treatment system suffice?
11. What is the estimated charge to each user for the onsite treatment system service if such service would be cost-effective?

Insofar as new installations are concerned, PL 92-500 and the regulations based on it impose no restrictions on types of sewage treatment systems. Systems such as septic tanks, holding tanks, package plant treatment systems, etc., are eligible for funding for state-approved and certified projects where minimum standards are met.

Cluster installations which serve two or more homes are eligible, although EPA may not be able to fund the construction of single-family onsite units. It should be emphasized, however, that alternatives in the facility plan should be analyzed and the most cost-effective options selected without regard to eligibility for federal funding.

A project must provide the most cost-effective method of waste treatment required to meet local conditions and must satisfy state and federal requirements, including water quality standards. Septic tank leach fields or other land disposal techniques must meet local, state and federal groundwater and public health criteria. Secondary treatment as defined by EPA regulations or some more stringent level required by water quality standards must be the minimum treatment provided if effluent is to be discharged to a stream or other body of water.

In order to receive federal assistance, a project must be owned, operated, monitored and maintained by a municipality as defined in the Public Law. The facility must be located on public property except where easements will suffice.

Vehicles and associated capital equipment required for servicing of the systems are also grant-eligible. Vehicles purchased under the grant must have as their *sole* purpose the transmission or transportation of liquid wastes from the collection point (*e.g.,* holding tanks) to the treatment facility. Neither general maintenance vehicles nor other types of vehicles are allowable for grant participation.

EPA Headquarters is taking a series of actions to make certain that onsite waste treatment systems are considered as an alternative in every small community where appropriate. Each state governor will be informed of the problem. EPA regional administrators are being asked to reject those grant requests where pertinent minimum action alternatives are not considered. Program requirements memoranda have been issued to require the originating and the review officials to consider small systems and to require the grantee's engineer to advise the public during the facility planning phase of what each family's estimated sewer costs will be if his recommendation is implemented. Our cost-effectiveness analysis guidelines have been supplemented with more detailed procedures for determining the cost-effective size of treatment works. These new proposed guidelines will soon be published in the *Federal Register* for public comments.

The EPA Office of Technology Transfer is preparing a seminar on small treatment systems to disseminate information on the state-of-the-art to interested officials and consultants. Seminars are planned to begin next spring and will be held in several of the federal regions.

To summarize, PL 92-500 (Section 201) grants may under certain conditions be spent for small wastewater treatment systems, servicing equipment, and residual waste disposal facilities. Many of the facility plans reviewed by Headquarters to date should have analyzed better the cost-effectiveness of utilizing such small facilities. Use of small facilities may in some cases reduce capital and O/M costs. Such a reduction is essential. Recent analyses of a large sample of completed facility plans for small communities resulted in the conclusion that many households may not be able to afford the facility recommended, and less expensive means of meeting treatment needs should be considered more carefully.

6

HUD'S RESPONSE TO THE HOUSING CRISIS—
CURRENT EXTENT OF AGENCY INVOLVEMENT

Orville G. Lee

    Housing and Urban Development
    Washington, D.C. 20410

I've passed by the Archives Building in the Federal Triangle between Constitution and Pennsylvania Avenues many times and I've noticed the engraving in stone, "The Past is Prologue." If HUD's response to the housing crisis over the last few years is "prologue," best we not lay too many plans.

Then, if we take a look at the national growth pattern over the past few years as "prologue," we note that if that trend continues toward a new urban-rural balance, the country is in real trouble. There are no oceans in rural America where we can quietly dump our sewage. It's a world of easily polluted small streams and rivers, already losing its groundwater.

Rural growth is on the increase. Metropolitan urban growth has leveled. Recent Census Bureau reports reveal that since 1970 metropolitan areas have grown less rapidly than the country. We note that service and manufacturing activities typically are locating in smaller towns and cities in counties adjacent to metropolitan areas, around state colleges and universities and in retirement climates. If this trend were to prove durable, it would constitute an end to the massive migration from rural hinterland to major urban centers that has continued virtually unabated since the early 1800s. The largest metropolitan areas registered the greatest decline. In a striking change, five of the eight largest areas experienced a net out-migration between 1970 and 1973. In contrast, these same eight areas had registered one-third of the nation's total growth over the course of the 1960s. Los Angeles showed the most dramatic turnabout. From 1970 through 1973 Los Angeles had a net loss of 119,000 residents, after having attracted a net gain of over 1.2 million new arrivals during the previous decade. Only Washington, D.C. has grown by as much as 1% since 1970.

Some observers attribute the post-1970 population shifts in large part to the increasing preference of some people for a rural or small town residential environment. If rural areas are experiencing a rebirth, the implications will be far-reaching.

We are all aware, of course, of the conflicting problems of growth patterns, interceptor lines and sewage capacity. Interceptor sewers servicing outlying areas have been blamed for inducing sprawl. Interceptor lines cross vacant lands and are said to induce infilling in by-passed areas.

Opposing questions are raised: Is sewage treatment capacity to be a self-fulfilling prophesy of growth needs in a community? Or are ample public facilities deliberately tied to local growth policy?

Many are now suggesting that rather than have engineers make *de facto* land use policy, the planning of sewage systems should conform to the desired growth pattern of the community.

Even though the EPA requires coordination of cost-effective sewage system design with land use planning, the program has been criticized in both directions. On the one hand, the EPA is being asked to make growth considerations a fundamental aspect of sewage treatment planning. On the other hand, there are fears that the EPA may become too involved in these questions.

Many rural communities are opposed to the introduction of sewer systems for the very reason that they are believed to stimulate higher density development. These communities are seeking alternative means of treating and managing wastes.

Septic systems are not adequate at higher densities and often not suited to certain soil types and geological areas. This is viewed by many rural areas as a means of ensuring the continued existence of the rural environment. This practice, of course, brings on the appearance of exclusionary zoning and all of its implications, as well as the impact of an increasing number of sewer moratoria around the country.

Sewer moratoria have been invoked in the name of environmental protection and the need to prevent untreated waste from contaminating waterways and water supplies.

The reasons for the moratoria, however, vary widely. In some areas, delays in funding and construction of treatment facilities are the root of the problem. In more areas, however, there is a deliberate attempt to use the absence of adequate facilities to gain time to reexamine community goals.

The center of the debate is the more fundamental need for the growth policy which respects regional population demands so as not to be exclusionary. Once growth policies are established, the best approaches can be provided for treating waste as well as protecting local drinking water supplies in the most efficient manner.

In order to fulfill proper growth policies more viable options are necessary. A variety of different onsite systems capable of solving the treatment problems under a great many political and physical conditions would at least preclude the sewage problems from being the whipping boy. If, indeed, the real problems are physical, new treatment options would open up whole new areas for growth planning.

The pressure is on for the development of this new tool. Onsite treatment devices are obviously the link to the solutions to environmental problems of water and sewer as well as to proper growth policy. For these reasons it is incumbent upon the federal government to encourage in some way the further research and development necessary to produce these new tools.

Now, you would naturally assume that either the Environmental Protection Agency or, even more logically, the Department of Housing and Urban Development should be in some way encouraging the development of such equipment. The industry is picking up the ball in some areas to some extent, but at the present rate of product development we'll be way behind because of further failures of thousands of disposal systems. The industry must have that further encouragement toward the development of a market.

HUD has a program plan under development and a small pilot project in the retrofitting of drainfields underway at the University of Washington. However, the funds to do a proper job are not readily available from the Building Technology research budget. HUD will be hard-pressed to come up with $100,000 this fiscal year and a quarter million in fiscal 1978. It will take at least 5 million dollars to make a decent start. An act of Congress with a special appropriation is really necessary to build an adequate program. Perhaps the amendments to PL 92-500 discussed by the keynote speaker could make this a possibility. A joint program between HUD and the EPA would be very appropriate.

The program can be viewed as one mainly of demonstration, similar to HUD's solar heating and cooling demonstration program. A modest program would have thrusts in several directions. First would be the development of the initial basic performance criteria. Device selection and testing would follow coincidental with demonstrations of all devices known—from composting devices such as the Clivus Multrum, through aerobic, anaerobic, recycling systems (including those for greywater), electric and chemical toilets, to even a few single-family total energy systems, and any other small-scale system that indicates potential. Following on the heels of the installation of the demonstrations would be an evaluation program to determine initial and continuing costs, maintenance requirements, water and energy needed for proper operation, overall results and, finally, user acceptance.

At the same time another thrust would be in conducting special related studies on such things as code acceptance, installation problems, marketing situations, costs, operation and maintenance problems, and other items that show up during the demonstrations.

Then in order to consider and demonstrate all systems, one more round of demonstrations would follow, to include those systems which were not ready on the first round, those which needed further research to qualify, or those systems developed because of the publicity on the program and the knowledge and realization of the national urgency.

Finally, manuals and information on the results must be written and made available to the industry and to the public. This would include more sophisticated performance criteria, operations and maintenance manuals, a continuing testing and labeling program, inspection procedures, cost and availability of equipment, administrative systems and procedures for local governments, soils studies, training programs, methods for determining the situations under which the various systems should be chosen for use, and the advantages and disadvantages of all systems tested and demonstrated.

At the same time we must continue with investigations into the retrofitting of failed systems. Since that problem is already out of hand, we need solutions in the worst way.

It is my opinion that we are facing a national crisis second only to the energy crisis. The increasing costs of ever larger municipal systems for both water and sewer and the resulting impact of these systems on the environment will prove to be more than local governments can stand.

The ideas and the knowledge of the equipment for solving these problems are available today.

I had mentioned HUD's septic system repair program. The text of the work statement follows.

STATEMENT OF WORK

I. BACKGROUND

Those concerned over the years with development of rural areas of the country have repeatedly encountered the continuing problem of disposing of household sewage safely and inoffensively on the premises. The problem is not new.

The decade after World War II produced the early suburban housing boom and the beginnings of scattered rural, nonfarm living. It also produced a demand for septic tank systems for disposal of sewage in many suburban areas. The U.S. Public Health Service carried on several empirical studies of the design and performance of these systems in an attempt to protect the public health as urban populations decentralized. Shortly thereafter, increased attention was given to the expansion of public wastewater collection and treatment systems, and research centered on the improvement of treatment methods for these large collection, treatment and disposal systems. Much of the impetus for this development occurred when it was discovered that massive numbers of homes served by septic tank systems for sewage disposal, were not an adequate long-run answer.

In the 1960s recreational second home development increased dramatically in many parts of the country. Many of these homes were small individual developments undertaken by the owner, but a few were very large integrated developments some of which contained several thousand homesites. Nearly all of these systems depended upon the traditional septic tank system for disposal of household waste.

The individual homesite waste disposal systems represent a significant sector, since they account for approximately one-third of the individual housing units in the United States. For this sector the primary alternative to the conventional central waterborne sewerage system has been the conventional septic tank system. Pennsylvania, for instance, with the largest rural population in the United States, finds that over two-thirds of its soils are unsuitable for conventional septic tank systems with subsurface disposal.

One of the more difficult problems experienced in Acquired Property Disposition in Region X is the single-family unit with a malfunctioning septic tank waste disposal system. Approximately 15% of the total single-family properties in their acquired property inventory are on septic tank systems. Thirty to forty percent are inoperative when acquired, making approximately 360 homes that require extensive rebuilding of the septic tank and disposal field to make them habitable. Under existing procedures the corrective measures involve rebuilding the system as originally installed using new soil hauled to the site. Typically, the failure repeats following two years of use.

It is essential to seek mechanisms to bring these housing units back into the market with a longer-term corrective action. Appropriate remedial action must be based on research data which are not now available for soil and weather conditions in Region X, and on experimental data to show the appropriateness of adequate design and maintenance.

The Washington State Health Department and the King County Health Officer are in full accord with the need for the data, and both will participate in obtaining the data through cooperation in obtaining it and by providing extra manpower support to the contractor. Both the State Health Department and the County Health Officer have stated their intentions for implementing the results of the experiments.

II. SCOPE OF WORK

A. Objective

A comprehensive study of onsite sewage disposal systems in Western Washington to be undertaken by the University of Washington Civil Engineering Department in four phases.

Phase 1: Design. This part of the study would focus on the conceptual and functional design of the two components of these systems. The major emphasis here would be on the provision of design modifications for existing systems that have failed and that are now under the jurisdiction of HUD. Concurrent with the latter stages of the design phase would be a construction and installation phase by HUD, which would provide the necessary test septic tanks and soil absorption fields needed for further study in the second phase.

Phase 2: Field Monitoring and Data Collection Phase. To characterize the performance of existing facilities experiencing difficulties and of modified experimental units the physical, chemical and biological water quality parameters listed below will be measured. Generally the influent to the septic tank, the effluent from the septic tank, and the effluent at the end of the soil absorption field will be monitored.

Water Quality Parameter

Dissolved Oxygen	Chloride
Biochemical Oxygen Demand	Sulfate
Chemical Oxygen Demand	Nitrogen
pH	Phosphorus
Suspended Solids	Coliform Bacteria
Total Solids	

Where possible, wastewater flows will be measured by metering the domestic water consumption at the residences. In certain instances it will be necessary for HUD to provide rental occupancy of housing contributing wastewater to the sewage disposal systems. It is also anticipated that HUD would apprise the renters of the study so that researchers may have freedom to carry out field observations and sampling observations.

Phase 3: Data Analysis and Report Preparation Phase. The data collected under Phase 2 will be statistically analyzed to determine the effect of study variables upon performance. It is expected that the results of the study will be used to develop reports and manuals which will contain valuable design, construction and maintenance information for septic tank and drainfield installations in Western Washington. Special attention will be directed toward the solution or mitigation of the problems existing at HUD residences. It is anticipated that quarterly progress reports will be prepared as well as a final report.

Phase 4: Seminar Phase. In order to more effectively disseminate the results of the research, Northwest Regional Seminars will be conducted. Efforts will be made to include several nationally known speakers, and it is hoped that the seminars will be essentially self-supporting.

INDIVIDUAL ONSITE WASTEWATER SYSTEMS

B. Specific Tasks

To accomplish these phases the University of Washington Department of Civil Engineering will perform the following tasks:

Task A: Develop state-of-the-art, "onsite" wastewater disposal.

Task A1—Review previous and current work, including the ongoing programs at the Universities of Wisconsin, Arkansas and Connecticut, and the Technical Study FHA 533, April 1964 by P. H. McGauhey and John H. Winneberger of the Sanitary Engineering Research Laboratory, University of California, Berkeley, California.

Task A2—Establish background data on the extent of improper onsite wastewater disposal in Western Washington.

Task A3—Describe state-of-the-art, including mounding and aerobic systems.

Task B: Select sites for field study.

Task B1—Develop site evaluation and selection criteria.

Task B2—Evaluate potential study sites.

Task B3—Recommend study sites and in consultation with HUD make site selection.

Task C: Construct and install new experimental septic tanks and drainfields.

Task C1—Select variables for evaluation.

Task C2—Layout detailed test plan.

Task C3—Develop and build test equipment.

Task C4—Conduct initial testing.

Task C5—Modify test plan as required.

Task D: Field monitoring and data collection.

Task D1—Develop field sampling procedures in consultation with HUD.

Task D2—Measure physical, chemical and biological characteristics in wastewater.

Task D3—Modify equipment and monitoring plan as required.

Task E: Determine influence of significant variables on septic tank and absorption field performance.

Task E1—Statistically analyze data.

Task E2—Determine effect of variables on system performance.

Task F: Prepare reports.

Task F1—Prepare quarterly progress reports.

Task F2—Prepare final report.

Task F3—Prepare design and operation manual.

Task G: Conduct Northwest Regional Seminar on onsite wastewater disposal.

Task G1—Plan seminar.

Task G2—Conduct seminar.

III. FIELD INSTALLATIONS

To provide information on various design and operational alternatives the following operations will be researched and tested:

1. Conventional systems (control unit)
2. Vented drainfield
3. Evapotranspiration system
4. Transpiration tests on evapotranspiration and conventional systems
5. Conventional system with alternating drainfields
6. Subcontract to the National Sanitation Foundation for air jet aerator study.

Because of the preponderance of Alderwood sandy loam it is recommended that this soil be the receiving area of choice. Several previous experiments have been done on this soil as background material for this research.

IV. PROCEDURE (LABORATORY)

Analyses of the wastewater samples will be made to determine the physical, chemical and biological characteristics of the wastewater.

Certain analyses will be made in the field for those parameters subject to change during storage. The other water quality tests will be carried out in the Sanitary Engineering Laboratory of the Civil Engineering Department.

V. PERSONNEL

Prof. D. A. Carlson
Prof. R. W. Seabloom
Prof. D. E. Spyridakis

VI. PERIOD OF STUDY

The total project is scheduled for 24 months. Calendar time will depend on the construction phase, but it is projected that the first three phases plus construction will occur in the initial 18- to 20-month period and the seminars would be in the final 4- to 6-month segment of the process. The time scale is shown below starting in March 1976.

March 1, 1976 June 1, 1976 March 1, 1977 September 1, 1977 February 28, 1978

Phase 1

Phase 2   Phase 3   Phase 4

VII. REPORTS

A. Quarterly Reports

Within 10 days after the close of each quarter of the contract period, the contractor shall submit to the Government Technical Representative a report evaluating the activity of the three-month period, which shall give information on the operation of the research, the modifications that have been made, problems that have arisen with recommended solutions, and actions to be taken during the next three-month period.

B. Final Report

At the conclusion of Phase 3, the contractor shall submit to the GTR a detailed final report. The final report shall present a detailed analysis and evaluation of the sewer systems used in the research with recommendations on implementing corrective action applicable to HUD, Washington State Department of Health, and the County Health Officers. This final report will be the basis for all affected agencies to promulgate regulations that assure future correction that will be the most permanent and cost-effective.

7
ALTERNATIVE METHODS OF REGULATING ONSITE DOMESTIC SEWERAGE SYSTEMS

David E. Stewart

Attorney, Dane County Regional Planning Commission
Madison, Wisconsin 53709

INTRODUCTION

Background

There is a great deal of interest in onsite sewage treatment and disposal. As a case in point, one need only recall that this is the third national conference on the subject held by the National Sanitation Foundation.

It is hardly necessary to reiterate the statistics about the number of onsite systems currently in use (or **misuse**) in the United States today. Surely we already know the magnitude of the usage, most of the problems with, and many of the technical aspects of, the various systems of onsite sewerage. Other speakers at this conference will expound further upon these points. Instead, I would like to discuss the onsite sewerage regulatory program most familiar to you—the one that you administer, comply with and/or live under. However, this is not possible, due both to the vast number of programs and my own limited knowledge of them. Thus it is necessary to speak of an assumed typical system and a general regulatory approach. I hope that you do not view this as a philosophical discussion of regulation. This discussion will conclude with both general suggestions on how to improve a regulatory program and suggestions for regulation of innovative systems. I invite you to consider your own regulatory program, and if it already has incorporated in it many of these suggestions, you probably have a better program than most.

Assumptions

The underlying premise of this paper assumes that the same basic administrative structure will be used to regulate both the conventional septic tank-soil absorption system, as well as the more innovative systems of onsite treatment and disposal. This assumption is warranted because in all likelihood the unit(s) of government, which is currently regulating conventional systems, will in the future be called upon to regulate the more innovative systems as they are developed and proven acceptable.

Additionally, as previously noted, there is great diversity of governmental units throughout the country involved in the regulation of onsite systems. This range of possible units includes local governmental units of general jurisdiction, such as towns, townships, charter townships, boroughs, villages, cities and counties, as well as a limited number of regional governmental units and a seemingly unlimited number of special purpose districts. It is assumed that appropriate specific units of government having adequate authority are available in all areas of the country and that the reader will substitute the appropriate unit of government, as needed, in this discussion.

This paper begins with an examination of the techniques available for regulation and then identifies three phases where regulation of onsite sewerage systems is required. The paper offers some general suggestions which might be used to improve any regulatory program and concludes by categorizing most innovative treatment and/or disposal systems within a matrix and suggests possible regulatory techniques for each category.

REGULATORY TECHNIQUES

There are only a limited number of techniques available to any regulating authority regardless of the subject matter of the regulatory program. Thus, a program to regulate onsite sewerage and suggestions to improve that regulation are, in fact, constrained to draw from the same "laundry list" of available regulatory techniques. Most of these techniques may be included in one of the following four subsets:

1. Direct controls over the onsite system itself
2. Controls upon the actors (*i.e.*, designers, installers, etc.)
3. General or indirect controls
4. Unfair or unlawful controls.

Prior to a brief discussion of the techniques, it must be noted that the regulating agency must have the authority to impose these controls. This authority might be in the form of statutory-enabling legislation enacted by the state legislature granting the regulatory agency (at either the state or local level of government) the necessary power to implement the regulatory program. This legislation may or may not be obligatory—requiring the designated agencies to regulate onsite sewerage. Further, the legislation might specifically designate the exact regulatory techniques which are to be used and in some cases might even prescribe the procedure to be used and might establish a fee structure.

Alternatively, certain possible governmental agencies possess adequate authority either via the state's constitution or by their general statutory grants of authority to implement a regulatory program consisting of many, if not most, of the available regulatory techniques. One example of such an agency might be the state agency responsible for protecting public health and/or water quality. As a second example, incorporated communities, cities and villages in many states have the authority to impose the controls needed for an adequate regulatory program. In many states, these "home rule" powers are also available to other units of local government, such as towns and counties.

Obviously then, it is necessary to first determine what unit of government is attempting to actually regulate onsite sewerage and then to carefully assess what authority it has. To be absolutely correct, this assessment must look at statutory and case law, as well as the state constitution, and often the assessment is made more difficult because more than one unit of government is involved in the program, *e.g.,* a state-county program.

Direct Controls

Direct controls may be thought of as those techniques which the regulatory agency imposes directly upon the onsite system itself. These controls can perhaps best be viewed in the chronological order of any onsite system.

Standards

Standards are the first direct control which could be imposed. Sizing, design and installation standards, as well as site requirements, have always been the basis of most regulatory programs. However, these standards for even the conventional system have been found to vary widely from one program to another.[1] This is especially so when comparing one state's program to another. Many individuals regulated by these programs and those involved in the regulatory programs have questioned this variance, since all programs are intended to have an identical purpose—the protection of public health.

Plan Review and Approval

Plan review and approval is a second direct control that frequently is included in a regulatory program, although by no means is this technique used in every program. For example, Wisconsin does not require the submission of plans for state approval of conventional onsite systems serving single families; however, pursuant to a recent amendment, the state now requires county review of such plans.[2]

Inspection

Inspection is a third technique which is used in most regulatory programs. The types of inspections which have been used range from the inspection of the proposed site prior to its approval for installation of an onsite system, to compliance inspections made after the system is completely installed. Included within this range are one or more inspections during construction. Note, the traditional "pre-cover-up" inspection would be included here.

Additionally, access to the property for the purpose of making the inspection has been the subject of several court cases. In summary, the United States Supreme Court has held that under certain circumstances inspections of property might be "searches" within the meaning of the fourth and fourteenth amendments of the U.S. Constitution and unless consented to can only be conducted or compelled under a search warrant. [See *Camara v. Municipal Court of San Francisco*, 387 U.S. 523 (1967) and companion case *See v. City of Seattle*, 387 U.S. 541 (1967).] Clearly, under these, holding nonconsenting inspections of a residence would require a search warrant. The courts have also extended this fourth amendment protection to include out-buildings and surrounding land (English law recognized this land as the curtilage or courtyard area). Thus it is likely that a warrant might be needed if permission cannot be obtained. For noncriminal proceedings many state statutes now prescribe the procedure for the issuance of an administrative warrant, since the standard search warrant procedure is generally not available.

Permits or Licenses

Permit or license issuance is a technique which is often used to regulate onsite sewerage. Generally, the program requires that a permit must be obtained prior to commencing installation of any onsite system. Clearly, the permit requirements may be varied for different types of onsite systems. In addition to the installation or construction permit, some programs require "use" or "occupancy" permits prior to occupancy of the residence served by the onsite system.

Also included under the rubric of this technique is the conditional permit—defined as one which is valid only until the occurrence of an event or the failure to comply with a requirement. The placard, stop work order or "red tag" might be used to show this occurrence or failure. The placard or "red tag" might be thought of as a permit (albeit negative) in its own right.

Monitoring or Surveillance

Monitoring or surveillance requirements are direct control techniques and are, in fact, similar to the inspection technique discussed earlier. The same constitutional constraints would apply to nonconsenting access to property as was discussed for inspections in general.

Controls Upon Actors

Onsite sewerage regulatory programs can obtain their primary objective, protection of public health, by means other than direct regulation of the onsite systems themselves. The most important method of so doing is to regulate those who act on the systems. Primarily the actors most likely to be regulated are the soil testers, designers and installers of onsite systems, as well as those who service or maintain the systems (*i.e.,* liquid waste pumpers/haulers).

Licensure

Licensure of qualified individuals has long been recognized as a legitimate function of the state under its police power. The United States Supreme Court described this power as follows:

> "The power of the state to provide for the general welfare of its people authorizes it to . . . secure them against the consequences of ignorance and incapacity, as well as of deception and fraud. As one means to this end it has been the practice of different states, from time immemorial to exact in many pursuits a certain degree of skill and learning upon which the community may confidentially rely, their possession being generally ascertained upon an examination of parties by competent persons, or inferred from a certificate to them in the form of a diploma or license from an institution." [*Dent v. West Virginia,* 129 U.S. 114, 122 (1889).]

As stated by the Supreme Court, the requisite degree of skill may be ascertained by an examination or inferred from a diploma or license from an institution.

Some of these regulatory programs have required that the designers of the onsite systems be licensed professional engineers, architects or plumbers. Additionally, some programs limit those who may actually install the systems to those who are licensed plumbers, architects, engineers, etc.

However, it is important to realize that the regulatory program is not constrained to rely on preexisting licensing programs, but may, in fact, provide a training and/or examination program and establish its own licensing program. For example, Wisconsin recently created a program to license those who perform the soils evaluations for the suitability of sites for onsite soil disposal systems.[3] Similarly, several agencies have apparently incorporated the licensure of the liquid waste haulers/pumpers into their regulatory programs.

Registration

Registration requirements are sometimes used as a regulatory technique. Generally the difference between this and licensure is that registration is often only a bookkeeping nondiscretionary listing of those who are performing certain functions. That is, registration might be nothing more than the keeping of an updated list of all those who have applied to the agency or otherwise indicated an interest in performing these functions.

One spinoff technique available to those agencies which impose licensure requirements upon some or all of those who perform actions upon onsite systems is to limit the issuance of permits solely to those who are properly licensed or registered.

General or Indirect Controls

Included here are controls which the regulatory agency seldom has the ability to influence or determine completely. Examples of these controls are zoning and land use policies. While the regulatory agency might have the enforcement or administrative responsibilities for zoning or other land use techniques, it is unlikely that the agency itself can adopt zoning land use ordinances. One exception to this would be the denial of the issuance of building permits until all onsite requirements have been complied with. It is possible that the same regulatory agency would process both permits.

A second example of an indirect control would be the existence of public policies in favor of or against certain regulated actions and these policies might aid or hinder the regulatory agency in the administration of its program.

Unfair or Unlawful Controls

The regulatory agency might seek to control certain portions of its onsite system regulatory program by establishing excessive fee requirements or by delay in processing applications. These techniques are not desirable control techniques and are only mentioned to point out that they do exist and have been used in the past.

THREE REGULATORY PHASES OF ONSITE SYSTEMS

The three phases where regulation of any onsite system is needed are the initial installation phase, the maintenance phase, and the failure phase. A good administrative program is one which adequately regulates all three of these phases.

First, the initial installation phase consists of proper siting and design requirements and proper construction of the onsite system. Through proper siting, installation, and design controls, the attendant problems of public health, chemical addition to the surface and groundwater, and economic hardship may be avoided. For this reason, a good regulatory program must impose siting and design requirements at this initial phase.

The second phase in the life of any onsite system is that of operation and maintenance. The problems of public health and chemical addition to the surface and groundwater may occur if the regulatory program lacks control over proper operation and maintenance. While there are very few operational or maintenance requirements for a septic system, some of the more innovative systems have more extensive requirements. Whether the system's operation and maintenance requirements are straightforward or elaborate, a good regulatory program should impose controls at this second phase in the life cycle.

The third phase occurs when a system fails. This phase involves both the detection of the failure and the necessary subsequent actions taken (repair or abandonment). This is the most difficult phase to regulate; however, the problems of public health, chemical addition to the wastewaters and economic and financial hardships may be attenuated or avoided by proper regulatory control at this phase.

SUGGESTED IMPROVEMENTS

Due to the variety of regulatory schemes used by the various states, the following suggested means of improving the regulation of onsite sewerage systems will not be applicable to all states and localities. Also, some of the suggestions may already be incorporated into the regulatory scheme in the jurisdiction of interest. Further, due to different state constitutional limitations and requirements, several of these suggestions may not be possible in all states. Also, many of these suggestions may require the enactment of enabling legislation.

These suggestions are discussed under the headings of the three phases in the life of an onsite system. Obviously, a suggested improvement may bring about improvements in more than one phase. In such situations, the suggestions are discussed in the phase where the most improvement might be effected. These suggestions deal first with the initial installation phase, and then operation and maintenance and finally the existing failing system phase.

Initial Installation

State Permit Program

It is suspected that many local health authorities are subjected to local political pressure to approve the installation of systems on unsuited sites. Aside from direct political pressure, some local authorities have reported that their boards of appeal have been subject to pressures and consistently override denials by permitting installation on sites thought to be unsuited by the local authority. To avoid this undue pressure, it is suggested that those states which do not presently have a state permit program should adopt one. The chance for direct political pressures at the state level should be less than the local level and the resources should be greater, in that the state either has or can employ soils or other experienced personnel to evaluate site suitability.

State Plan Review and State Standards

As an alternative, it is suggested that states adopt a mandatory plan review of all the onsite systems approved at the local level. This state review process would be conducted by the appropriate state authority and would prevent the use of systems on improper sites by countermanding local approval when required. As an alternative to plan review, it is suggested that the state enact a mandatory review of all local sanitary programs; and when a local program is found to be deficient, the state should impose a state program until the locality brings its program up to standards. The state would have to establish minimum standards for local programs including enforcement practices, staff requirements, employment practices, siting and installation inspection requirements, etc.

Also, these standards could even set out design and siting requirements for onsite systems.

Uniform Citation and Complaint

States and localities not having a method of issuing citations for sanitary ordinance or code violations are urged to adopt such a system. The citation system is currently being used by building inspectors in several major American cities to "ticket" owners of buildings which violate local codes.[4] Essentially the uniform citation is similar to a traffic complaint and the violator of the sanitary ordinance (homeowner or system installer) signs the citation and agrees to appear in court to enter his plea. This system cuts down on enforcement delays and permits the local or state health authority to issue citations for violations as he sees them at the time of the violation. This system is equally applicable to violations in the other two regulatory phases.

Small Claims Courts

Many states have small claims courts for cases involving an amount less than a given number of dollars. Usually, these courts allow an abbreviated, less formal procedure generally using printed forms. When seeking only fines or forfeitures, state and local authorities are urged to consider using these small claims courts to prosecute initial installation violations, as well as all other sanitary ordinance and code violations. Generally, there is a smaller backlog of cases in these courts than in courts of general jurisdiction; thus the enforcement of sanitary violations can be accelerated. Note that special enabling language might be required in some states.

Civil Service Status

Many local regulatory officials and some state officials serve at the pleasure of those who appointed them to their jobs. It is assumed that this lack of job security has hindered vigorous application and enforcement of initial installation requirements as well as enforcement of the other phases of regulation. To give them the necessary job security to do a vigorous job of enforcing the sanitary requirements, especially the crucial siting requirements, local and state agencies are urged to seek a civil service program for these officials.

Operation and Maintenance

Septic Tank Maintenance Permit

This phase of onsite system regulation is often the most overlooked. State or local authorities are urged to adopt a maintenance permit program to assure that septic tanks will be inspected once in a given number of years (1, 2 or 3) and that the septage will be pumped when necessary. The program would require the licensing of the pumpers. The homeowner would be mailed a maintenance permit form and would be given, say, 60 days to have any licensed pumper inspect and, if required, pump his tank. The pumper would sign one portion of the homeowner's permit thereby certifying that he inspected (and pumped) the tank. The authorities would then have on file a certified statement that the tank was inspected on a given date. Then just prior to the expiration of the 1-, 2- or 3-year period, a similar card would be sent to the homeowner to renew the permit by repeating the process. Of course, it would be unlawful for any owner to use his system unless he held a valid permit. Also, this maintenance program could be modified so as to apply to other more innovative onsite systems.

Conditional Sanitary Permit

As an alternative to the maintenance permit program, state and local authorities which issue sanitary permits for onsite systems can make these permits valid subject to the condition that inspection and pumping (if necessary) be performed every 1, 2 or 3 years. The enabling legislation or ordinances would have to be worded to make it unlawful for a system owner to use his system unless he had a valid sanitary permit and the permit would be valid only if the necessary inspections (and pumping) had been performed.

Location Filing Requirement

Many state and local authorities already require the filing of a plan of the proposed (or built) system. For those that do not, they are urged to impose the requirement that each system owner file an "as built" plan of his system, clearly referencing the location of the system manholes. Such a plan is invaluable when it becomes necessary to inspect or service the onsite system. It has been noted that many owners do not know the location of their systems and obviously this makes maintenance difficult. In an attempt to improve this phase of regulation, state and/or local authorities are urged to adopt this filing requirement and to establish a file for these plans and index them by street address, name of original owner, installer and perhaps legal description.

Failing Systems

Sanitary Surveys

Detection of the failing system is one of the most important aspects of this final regulatory phase. If state and local authorities do not already have the authority and funding to perform sanitary surveys, they are urged to obtain them. The large staff commitment and the expense of such surveys are recognized, but these are justified as surveys are the most thorough method of determining which existing systems are failing.

Violation as an Encumbrance

In many states, the effect of a sanitary code violation on the title to the property is unclear. In an effort to give notice to potential buyers of land containing sanitary violations, especially existing failing systems, states and localities are urged to pass legislation which makes the violation an encumbrance on the title. Such an encumbrance will put buyers on notice of the violation and will probably lower the price of the property since the seller does not have a clear title.

Pre-Sale Inspection

An alternative to the encumbrance would be to require an inspection of the onsite system prior to the sale of the property. Legislation or ordinances could be worded to either require the correction of all violations before permitting a sale or to encumber the title.

Abatement Costs

Many regulatory authorities have the authority, under certain conditions, to enter onto private property and to abate or correct violations—usually a failing onsite system. Those regulatory agencies which lack this authority should lobby to get this power. Further, however, it is necessary that the enabling legislation specifically provide that the cost of the work may be added as a tax on the lands upon which the violation occurred. Also, the agency should be given the authority to contract to have this work performed.

REGULATION OF INNOVATIVE SYSTEMS

"Innovative" for the purpose of this paper can best be defined by stating that it is any system other than the conventional septic tank-soil absorption field. The matrix shown in Figure 1 attempts to categorize all the possible types of innovative systems. Any system may be thought of as having a treatment and a disposal aspect. The possible combinations of treatment and disposal are not limited as can be seen in the matrix. Essentially, any system, whether available today or yet to be developed and/or proven, will have either a conventional, innovative or no treatment method coupled with conventional, innovative or no (containment) disposal.

			Disposal			
				Innovative		
		No (Containment)	Conventional (Subsurface Soil)	Soil	Surface Water Discharge	E-T
Treatment	No	1. Holding tanks 2. Recycle toilets (non-water carriage) 3. Self-contained toilets	1. Cesspool	Unlikely	Unlawful	Unlikely
	Conventional (Septic Tank)	Unlikely	1. Septic tank-soil absorption field	1. Mound systems 2. Fill systems	Probably unlawful	1. E-T beds
	Innovative	1. Composting toilets	1. Mechanical aeration-soil absorption field	1. Mechanical aeration-mound system	1. Sand filters 2. Physical-chemical 3. Lagoons 4. Mechanical aeration 5. Others	1. Not known at this time

Figure 1. Matrix of anticipated possible treatment and disposal combinations of onsite sewerage systems.

Figure 1 also contains a listing in each category of examples of the types of systems suggested by each category. As can be seen, several of the categories are unrealistic or unlawful and will not be considered further. Comments about the regulatory programs for the remaining categories follow.

First, the regulatory program for the holding tanks and toilets (no or innovative treatment—no disposal) would not require much in addition to the program for the conventional system. The eventual disposal from the holding tanks and toilets poses the most serious potential threat to public health. Licensure or registration of those who service these systems would probably be a useful regulatory technique. As a requirement for licensure, the licensed personnel should in addition be required to maintain detailed records of service and disposal locations and supply copies to the regulatory agency.

Second, the regulatory program for conventional disposal systems (regardless of the method of treatment) is not expected to require much in addition to a good conventional program. Perhaps innovative treatment with conventional disposal could demand additional inspections of the treatment device and should include some form of guarantee that the treatment system will be properly maintained. For example, permits should be conditioned upon the existence of a valid paid service contract with either the distributor/supplier or with a licensed treatment plant operator. Alternatively, maintenance could be performed by a special purpose district created to own and/or operate all such systems within a given area. This assurance of proper operation and maintenance is especially warranted if the conventional absorption field size is reduced because of the assumed improved treatment provided by the innovative treatment system.

Third, innovative soil disposal methods do require an increased regulatory effort. This results because the systems are designed for use on sites that are less suited for onsite sewerage. That is, the public health risks are likely increased if there are any malfunctions. Regulatory techniques such as a more thorough site evaluation, an increased number of inspections during construction, agency plan review and approval and a program of monitoring (because of the increased public health risk) should be imposed. Again, these functions might be performed by the regulatory agency itself, or by a special purpose district.

Fourth, any surface water discharge will have to comply with federal (PL 92-500) and state water quality standards. Each discharge will be required to obtain a pollution discharge permit (NPDES) and meet federally set discharge standards (currently secondary treatment requirements). Also, included in these federal requirements are monitoring and reporting minimums which must be met by each discharger. Thus, the regulatory program for this category of treatment-disposal will in a large part be determined by the federal requirements under PL 92-500. Of course the regulatory agency may employ additional regulatory techniques such as plan review, and the requirement of the use of licensed treatment plant operators to inspect and maintain the innovative treatment facilities. Again these functions could be performed by a special purpose district.

Finally, the innovative disposal technique involving evapotranspiration might include in its regulatory program techniques such as plan review and increased inspections during construction.

In conclusion, some of the categories of treatment and disposal methods may require increased regulation because of the potential increased risk to public health. The regulatory techniques discussed earlier should be adequate to protect against this increased risk.

REFERENCES

1. Plews, G. D. "The Adequacy and Uniformity of Regulations for Onsite Wastewater Disposal—A State Viewpoint," *Proc. Second National Conference on Individual Onsite Wastewater Systems*, NSF, Ann Arbor, Michigan (1975).
2. Wisconsin Administrative Code, Section H62.20, State Documents Office, Madison, Wisconsin (1975).
3. Wisconsin Administrative Code, Chapter H64, State Documents Office, Madison, Wisconsin (1974).
4. National Institute of Municipal Law Officers, Municipal Law Review, 33:342 (1970).

8

ONSITE WASTEWATER DISPOSAL: A LOCAL GOVERNMENT DILEMMA

Paul Pate

    Director, Jefferson County Department of Health
    Birmingham, Alabama 35202

Every community, whether urban or suburban, feels a driving need to grow both physically and economically, as well as to prosper as a living, thriving, economically healthy entity. Economic and political considerations, civic pride, along with population growth, all propel them in this direction.

At the same time, the need to protect each local area from environmental degradation, and to keep its natural assets such as waterways alive, healthy and viable, is not only a justifiable concern of its citizens but also is supported by a variety of state and federal regulations.

These two objectives inevitably rub against each other at some point and even occasionally assume the appearance of cross purposes. This circumstance is especially evident in the problem of acceptable wastewater disposal, a problem which is burgeoning in complexity on a nationwide scale and pressuring local governmental units to make solutions materialize where no solutions seem to exist.

The present trend toward progressively more stringent regulatory control over onsite wastewater disposal, evident in governmental agencies at several levels, lies near the heart of this dilemma which has such potential for adverse impact on community growth. This trend must be examined in its confluence with the trend toward simultaneously escalating restrictions on construction of other types of wastewater disposal systems which have been legislated at both the state and federal levels.

A thoughtful reexamination of the full potential for utilization of onsite systems is virtually demanded because of legal, technological and financial roadblocks to total reliance on conventional sewer and treatment systems.

Among these roadblocks are progressively stringent federal and state legislation restricting wastewater discharge permits and regulating the quality of wastewater, as well as the number and extent of federal studies (such as 201 and 208 and Environment Impact Studies, scenic and wild river studies) required of governmental agencies regulating municipal and industrial wastewater.

While many control agencies are seeking to eliminate onsite disposal altogether, let us take a closer look at the squeeze in which local governments find themselves. They are under pressure from federal and state agencies, local planning agencies, and environmentalist groups to provide treatment and waste removal sufficient to satisfy the requirements and/or aims of each. Federal agencies require strenuous effluent limitations on discharges of pollutants into receiving streams, as do state regulatory agencies. Environmentalists are justifiably concerned for the future of our streams and are seeking to halt their degradation and provide some chance of recovery in the distant but foreseeable future. In the struggle to satisfy all these demands, economic growth can grind to a standstill.

Coupled with rapidly increasing construction costs, these trends can impede the home building industry sufficiently to produce an economic crisis.

The degree to which growth and construction for the past 20 years has relied upon the acceptability of onsite disposal systems places an additional financial burden on the proposition that onsite disposal should be minimized or eliminated. The community which seeks to eliminate onsite disposal altogether must be prepared to deal with a large aggregation of such burdens. Among those which must be faced are:

1. The cost of replacing all existing onsite systems with sanitary sewers.
2. The cost of expanding existing interceptor sewer systems and treatment facilities to meet the demands of the additional loadings stemming from replacement of onsite systems.
3. The cost of rehabilitating old sewers to eliminate gross infiltration.
4. The cost of upgrading existing primary and secondary treatment facilities to meet the demands of increased restrictions on effluent discharges, while simultaneously increasing their capacity adequately to allow for a growth factor.
5. The growing complexity of sludge removal problems.
6. The trend toward cut off of federal fundings to local agencies for assistance in construction of sewers and treatment plants.
7. The likelihood that stresses on energy resources will continue to escalate the cost of energy for operating treatment facilities, which already approximate $40 per day per million gallons of wastewater treated.

In addition, it should be remembered that our present concepts of treatment costs are based only on primary and secondary treatment. In many areas (such as Jefferson County, Alabama) in which minimum and seven-day low flows are considerably less than wastewater flows, trends for the readily foreseeable future will demand third- and fourth-stage treatment. If and when this occurs, costs will be unlimited.

The magnitude of the financial snarl which could develop from total rejection of onsite wastewater disposal can be illustrated in the following figures which describe the problem in Jefferson County, Alabama.

In Jefferson County, as in many other metropolitan communities, the potential for metropolitan growth is tremendous if land area is the only criterion.

This county has a land area totaling 1118 square miles. Of this amount, 182 square miles are served by sewers and 164 square miles by onsite wastewater systems, leaving 772 square miles undeveloped and available for future use.

The county presently has 11 major treatment plants, approximately 4.6 million (4,655,864) linear feet of sanitary sewers, and about three-quarter million (763,522) linear feet of interceptor sewers. Extending sanitary sewers to the municipalities and unincorporated areas presently served by onsite systems would require an additional 4 million (4,037,365) linear feet of sanitary sewers and another 350,000 feet of interceptors. In other words, sanitary sewer lines would approximately double and interceptor sewers would be increased by approximately 50%.

To translate this situation into dollars: In excess of $54 million worth ($54,308,623) of work in upgrading primary and secondary treatment facilities is currently under contract. An additional $71.5 million is needed just to upgrade the balance of the system adequately for present loading of the system. It is estimated that the cost of replacing all onsite systems with sanitary sewers would be approximately $236 million. If this amount were added to the $71.5 million in improvements already needed but not contracted for on the present system, this would translate into a cost of $557 per capita or $2230 per family for the taxpayers of Jefferson County. If this seems an insurmountable obstacle to local governments, consider the alternative of direct assessment of costs of new sewer installation to the residents to be served by them. Onsite systems are being used by a population of 160,000, or roughly 40,000 families. Their assessments would approximate $1479 per capita or $5917 per family.

It should also be remembered that these figures do not include estimates for the expansion cost of treatment facilities to receive additional loadings from all areas if onsite systems are rejected for future growth.

If the barriers to total reliance on sanitary sewer systems and conventional treatment are formidable—and they are—it must be remembered that there are also barriers (in addition to regulatory trends) to optimum utilization of onsite systems. Most of these concern lack of knowledge. There is a real and pressing need for new and expanded research programs to provide answers for questions which present programs have inadequately answered. Among the questions which must be addressed are the following:

1. What environmental hazards are associated with use of onsite systems: Sludge volume reduction and groundwater contamination from viruses, bacteria, nutrients and trace metals?
2. What geological formations either support or remove these contaminants?
3. How do varying rock formations affect the ability to cleanse wastewaters injected at shallow depths?
4. How do onsite systems and collector systems relate or contrast with respect to reduction of sludges, viral and bacterial loadings, and nutrient loadings not only to groundwaters but also to recharging areas that supply streams and lakes?

While research efforts into such questions as these must be accelerated, we must also continue to search for innovative approaches to wastewater disposal problems that involve both conventional and onsite systems in some combination that can be deemed feasible, environmentally safe, and economically sound.

The effectiveness of the utilization of onsite systems can be demonstrated by Figures 1, 2 and 3, and Tables I and II illustrating the growth of areas served by them and analyzing the sucesses and failures of the systems.

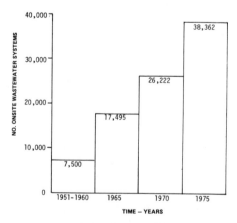

Figure 1. Onsite wastewater systems installed (accumulated), 1951-1975 Jefferson County, Alabama.

INDIVIDUAL ONSITE WASTEWATER SYSTEMS 71

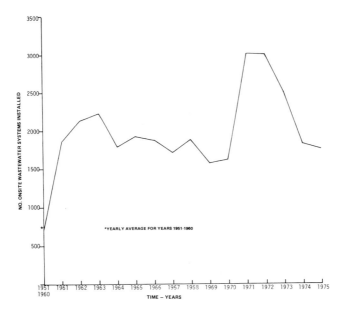

Figure 2. Onsite wastewater systems installed, 1951-1975, Jefferson County, Alabama (line graph)

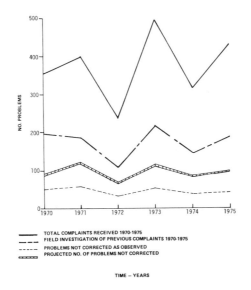

Figure 3. Onsite wastewater investigations and analysis of problem systems, 1970-1975 Jefferson County, Alabama

Table I. Complaint Analysis: Onsite Wastewater Systems (1970-1975)

	1970	1971	1972	1973	1974	1975	Total
Number of complaints received	356	400	238	495	316	430	2235
Number of complaints unable to locate or no one home	160	215	129	278	194	236	1212
Number at home and information obtained	196	185	109	217	142	194	1043
Justified complaints	150	129	94	198	122	164	857
Complaints not justified	46	56	15	19	20	30	186
Justified complaints not abated according to sanitarian	50	56	32	52	38	41	269
Average age of house since first complaint (years)	15	12.5	18.5	12	16	23	16.1
Projected number of justified complaints	90.8	121.1	69.9	118.6	84.6	90.9	576.4
Percentage of total complaints not abated	25.5	30.3	29.4	24.0	26.8	21.3	25.8

Table II. Onsite Wastewater Investigations and Analysis of Problem Systems (1970-1975)

Year	Number of Septic Tanks Approved	Number of Complaints	Justified Complaints	
			Number	% of Total Septic Tanks Installed
1951-1960	7,500	209	54	0.72
1961-1965	9,995	872	225	2.25
1966-1970	8,727	1,159	299	3.43
1971-1975[a]	12,140	1,879	485	4.00
Total	38,362	4,119	1,063	2.77

[a]Based on 25.8% complaints justified as determined by an average of actual complaints justified over a 6-year period (1970-1975) as developed from complaint analysis.

ACKNOWLEDGMENTS

I would like to recognize and acknowledge the assistance of the staff of the Jefferson County Health Department's Bureau of Environmental Health in compiling the data necessary for this preparation.

9

AN OVERVIEW OF DISPOSAL OPTIONS: THE ONTARIO PROGRAM

D. M. C. Saunders

    Ministry of the Environment
    Toronto, Ontario M4V 1P5

INTRODUCTION

A discussion of the Ontario program of onsite sewage disposal options is obviously broad in scope, and as an "overview" full detail is not possible. The types of sewage disposal systems which are authorized for use in Ontario are stressed, rather than the regulatory aspects. However, frequent reference will be made to the act, regulations, policies and guidelines that, in effect, authorize the use of onsite disposal systems and set the standards which must be followed.

For clarification a brief outline is presented covering the authorities that set the standards for onsite sewage disposal, and those that implement the programs in the field, as there will be reference to them in the remainder of this paper.

LEGISLATION, REGULATION AND ADMINISTRATION

Ontario's Environmental Protection Act, Part VII, deals with sewage systems other than those which discharge to receiving bodies of water. The Regulation (Ontario Regulation 229/74) indicates that such systems must be contained on the property on which the building or buildings served are located. The exception to this is the system which pumps out septic and holding tanks and transports sewage offsite for disposal elsewhere.

The Act covers such legislation as the requirements for system approvals, permits, licenses, the powers of the approving authorities and the right of appeal of the applicant.

The Regulation classifies sewage systems by a number classification and is the medium for setting standards for each type of system and such details as fees to be charged. The provisions of the Regulation are mandatory although an approving authority is permitted to make exceptions under the Act where these exceptions are specific only to a particular application.

The policies, procedures and guidelines of the Ministry are contained in a field handbook, soon to become a manual, which is essentially a reference and guide for those administering the program. While the Act and the Regulations are law, the handbook sets out methods to be followed, forms and procedures to adopt, and gives technical and administrative guidance in all matters relative to the implementation of the program.

The Provincial Government, represented by the Minister of the Environment, is responsible for the overall program and for the Act, Regulation and guidelines. Throughout the province, except in one area which is essentially a recreational lake area, in the unorganized territories, and in the federal properties and the provincial parks, the program is administered by the municipalities at the regional government and county level, and the implementation is done by the staff of the Medical Officer of Health of the Municipality. This delegation is arranged by formal agreements. As an indication, there are 43 such health units administering the onsite sewage program.

FACTORS AFFECTING SEWAGE DISPOSAL PRACTICES

General

Ontario is not by any means a homogeneous province. There is great variety in terrain, climate, population density, property usage, and types of sewage systems. This is of course true to some extent in all of the United States, but the differences are very marked in Ontario as such an extensive area is covered.

Terrain

Included under this heading are the differences in the typical soils encountered, land drainage, prevalence of rock (fissured or otherwise), groundwater and slopes.

Climate

Climate affects the types and standards of sewage systems primarily because, in all parts of the province, there are several months of winter weather with subfreezing temperatures. The more northerly parts experience a long cold winter.

Population Density and Property Usage

The combination of terrain and climate have a marked effect on population distribution, on the type of residence that is served by onsite sewage systems and, indeed, on the usage of the residence by the occupants. For example, parts of Ontario in which there are many lakes suited for recreational purposes, and which are reasonably close to the centers of population, are frequented by cottagers in all seasons. These areas are predominantly rocky, either precambrian or limestone, and have limited soil cover. One factor becoming increasingly common in areas of this type that are within commuting distance of places of employment is the tendency for winterization and year-round, if not continuous occupancy. Retirement to areas previously considered as summer cottage areas is also increasing. Better transportation and the snowmobile have made this more practical, but it presents many problems for sewage disposal and tends to shape policies that cater to the possible future use and development of the home at the expense of those who may genuinely wish only a summer weekend retreat in which they can enjoy a quiet simple life and which could be adequately served by relatively primitive sewage disposal methods.

Sewage Systems

There are some 750,000 private sewage systems in Ontario with about 250,000 of these being primarily associated with recreational areas. They cover all standards of design permitted over the years by the approving authorities, or deemed appropriate by contractors or owners before the authorities laid down standards. While many of the old systems have given way to sewers as they become available, or have been improved or enlarged, many are still as they were originally installed and are functioning satisfactorily.

The long service that these systems have given, despite the fact that they may be grossly substandard according to present regulations and serve residences that have been substantially modernized since their original construction, is a point not often considered. We naturally hear much more about the failures than the successes, but they should not be forgotten;

because of this, authorities sometimes have difficulty explaining the imposition of new and ever tougher standards.

As in other areas, onsite systems serve all types of occupancies. While individual residences are the most common, there are many systems serving mobile home parks, tent and trailer parks, condominiums, motels, service stations, small industries, shopping plazas and rural nursing homes, to mention some typical installations.

The systems as categorized in the Regulation are as follows:

Class 1—Privies and Toilets. These are systems disposing of human waste only and include all types of privy, chemical and recirculating toilets, incinerating toilets, and composting toilets. The latter toilets are ones developed in Sweden and recently introduced to the Ontario market. They include the Mul-Toa, marketed in Canada under the name "Humus" toilet, and the Clivus Multrum. In addition to human waste, they may receive some organic waste, such as plate scrapings or vegetable matter, to assist in the degradation process of the human waste.

Class 2—Leaching Pits. This is the conventional leaching pit or soak pit, and is permitted for the disposal of the waterborne waste other than human waste, frequently referred to as greywater. While our regulations do not specify limits to the use of leaching pits, our policies do; and their use is not permitted in conjunction with new building construction if the water system is to be pressurized, or is likely to be pressurized in the future. Leaching pits are commonly found in cottage country and may be used to abate pollution problems that occur in such areas where an inadequate septic system is overstressed and is causing a public health nuisance, and where splitting the sewage disposal into a Class 1 and a Class 2 system may provide a solution.

Class 3—Cesspool. This is a covered soak pit constructed in the soil and is to be used only for the disposal of effluent from a Class 1 system, such as a pail privy, a vault privy, a chemical toilet, etc.

Class 4—Septic Tank System. This is the conventional septic tank system providing a solution to the disposal of all sewage from a property other than sewage of a quality not suited to such treatment. The design is based on the daily sewage flow and the soil conditions at the site where the leaching bed or absorption field will be located. A basic requirement for the leaching bed is that there must be a minimum of 3 feet of soil of percolation time not greater than 60 minutes per inch below the bottom of the absorption trench and above rock, high water table, or soils of percolation time exceeding 60 minutes per inch. The bottom of the absorption trench is about 2 feet below the surface resulting in an overall

INDIVIDUAL ONSITE WASTEWATER SYSTEMS

depth of bed of 5 feet. Systems up to the size serving a 5-bedroom house are covered by tables providing a minimum size for the septic tank and minimum length of distribution pipe (the name we give to the clay tile or perforated pipe used in the absorption trenches). For uses other than the residences that are covered in the tables, there is a formula for calculating the size of the tank and another for determining the total length of distribution pipe required in relation to the soil. Some essential parts of our regulations are:

- Tanks for daily sewage flows of up to 3600 liters are to have a minimum working capacity of two times the daily flow (or as required by tables).
- For daily flows over 3600 liters, the size is based on three-fourths daily flow plus 4500 liters.
- Tanks must have two compartments, not including the pump or siphon chamber; the first and second compartment should be about 2 to 1 in volume.
- In leaching beds of the absorption trench type, the trenches are to be a minimum of 6 feet on centers.
- Pipe must be a minimum 3 in. in diameter except for beds of over 500 lineal feet or beds dosed by syphons or pumps, in which case the minimum size is 4 in. diameter.
- There must be 3 feet of acceptable soil between the bottom of the absorption trench and any rock or unacceptable soil strata or high water table.
- Without changing the general layout, beds may be raised up or mounded where rock, unacceptable soils or high water table exist to ensure that the 3 feet separation is attained. In such cases the regulations add restrictions such as minimum side slopes, increased clearances from surface waters, etc. Where rock or unacceptable soil is at shallow depth, the raising of a bed must not result in a surface breakout of sewage that is forced to move laterally after it has trickled downward through the bed and meets the resistance of the "impermeable" strata. Guidance is given to inspectors to ensure that the contact area between the acceptable material in the bed and the unacceptable soil below is sufficient to promote absorption of the bed effluent by the underlying soil or by a minimum depth of pervious upper soils surrounding the bed in which plant growth can be sustained. Although the total depth of beds is about 5 feet (2 feet above and 3 feet below the bottom of the absorption trench), proposals are not rejected for septic systems where there is less than 5 feet of acceptable soil available naturally, as would appear to be the case in some other jurisdictions. An effort is made to meet the demands of development in such areas by permitting raised beds and hopefully setting standards which will ensure that they function properly.
- Area beds or sand filters are not at present permitted with septic systems, although trials at an experimental site have indicated that such

beds, using a filter medium having an effective size of 0.25-1.0 mm and a uniformity coefficient not exceeding 4, will function over a prolonged period with loadings up to 1.5 gallons per square foot.

- Beds must be dosed with a pump or siphon if they have over 500 lineal feet of distribution pipe. The dose must be about three-fourths the volume of the distribution pipe, delivered to the bed in not more than 15 minutes.
- One line or run of distribution pipe cannot exceed 60 feet if the bed is gravity-fed, or 100 feet if dosed.
- Distances to surface waters or wells with a water-tight casing to at least 20 feet is to be not less than 50 feet and not less than 100 feet to all other wells and springs.
- Other clearances from the outer pipes in a leaching bed, all measured horizontally, are 10 feet to a lot line, 25 feet or 10 feet to a structure, depending on the elevation of the pipes in relation to the lowest floor in the structure.
- Septic tanks are 5 feet, 10 feet, and 50 feet from structures, lot lines, surface water, and wells respectively.
- Systems cannot be installed on slopes in excess of 25%.

Class 5–Holding Tanks. The minimum size permitted is 9000 liters (2000 I.G.), and there are certain regulatory requirements for the tanks which are used in such systems. As they are dependent on being pumped out by a sewage hauler, holding tanks are considered somewhat of a last resort for private sewage disposal. Except as an answer to a temporary requirement, they are not generally approved in conjunction with new construction. Their use is more common as a solution to a malfunctioning disposal system where septic or aerobic systems cannot be installed.

Class 6–Proprietary Aerobic Treatment Systems. These are the aerobic, individual home plants. Several makes have been introduced into the province. A particular model can obtain "acceptance for use in Ontario" providing:

- The manufacturer demonstrates through test results that its operation will not be detrimental to the environment or public health.
- It is not complex in operation and therefore does not require a degree of servicing not obtained with private systems. Class 6 systems cannot be operated without a continuous service contract, either with the manufacturer, his agent, or a licensed service contractor.
- The leaching bed is either a filter type specifically designed to handle the maximum daily output of the plant, or is of the conventional absorption trench design. In the latter case we allow some reduction in trench requirements in relation to septic systems when the trenches are in soils of high permeability, but bring them progressively into line so that for soils of percolation time of 45 minutes per inch or higher, they are virtually the same.

Class 7—Sewage Hauler. This is the classification given to the system that pumps out septic or holding tanks and transports the sewage for disposal elsewhere. Some temporary storage may be included. The operators of such systems must be licensed to be "in the business" and must, in addition, obtain specific certificates of approval for each operation and disposal method.

APPROVALS

The approval process involves the health units and the ministry in various ways with the agencies who are the authorities in the land use and housing fields. Each municipality should have an official plan approved by the Ministry of Housing on the strength of which municipal bylaws are enacted. The proposed plans and any amendments thereto are reviewed, and any overall observations made relating to areas on the plan where onsite sewage is proposed. Another task relates to the circulation of plans of subdivision or applications for land severance which are passed to the health units for review if onsite sewage disposal is proposed. The inspection of land then carried out is to determine if each proposed lot will be suited to onsite sewage disposal. The intention is to prevent the subdivision of land into lots which may subsequently be assessed as inadequate for the installation of a sewage system when the eventual owner proposes to build and submits his application for approval. Provision for a septic system for a 3-bedroom house, with or without improvements to the natural lot conditions, is the basis of assessment unless the proposed construction is known.

With the exception of Class 1 systems, all systems require a certificate of approval from the health unit before they are constructed, installed, extended or altered; and, in fact, building permit offices require such a certificate before they will issue the building permit for the building which is to be served by the system. Subsequent to completion of the authorized work (at the "open trench" stage for a new system) the system is checked by the health unit and, if the work has been completed according to the terms of the certificate of approval and good workmanship, it is accepted and a use permit is issued. An owner cannot operate a system without such a permit.

PROBLEMS AND DISCUSSION

The present approach and standards for onsite sewage disposal have been outlined, but in the application of these standards there are still unsolved problems. Sewage systems are, after all, a service to a building or development; and, as the nature of development changes in response to the changing public demands, there is a need not only to adapt techniques to these changes, but to continue investigations and to try out new ideas that may provide better solutions. Test establishments such as the National Sanitation Foundation perform a very important role in this area. The regulatory and program-implementing agencies must, of course, consider not only the theory and the technical adequacy of new equipment as tested, but also the practical application of such new ideas. As mentioned before, an unfortunate truth is that what is good for one part of a jurisdiction may not be practical in another.

Hopefully, the standards and guidelines to be applied in the field are practical as well as being protective of the environment and public health, and they should be developed with input from the "front line" agencies to achieve this end. In applying the standards to any specific application for onsite sewage disposal, it is necessary to visit the site and assess the subsurface conditions of soil, rock and water tables to the required depth, and to determine any parts of the lot where slopes are restrictive to sewage system construction. The surface drainage pattern must be studied, and a knowledge of the groundwater conditions obtained from the standpoint of any subsurface movement which would be adverse to the system, and any detrimental effects that the system may have on an aquifer. This assessment is no easy task. It involves many considerations and a knowledge obtained from experience. It would be ideal if all inspectors were experts on all points that must be assessed, but this is never so. Problems of prime concern are the assessment of a percolation time for the soil, either for use in tables or to determine an acceptable loading rate, and the determination of the water table elevation to be used in design.

Assessment of Percolation Time

Because the percolation time of the soil is used in all our tables and our formula for computing the distribution pipe requirements or an acceptable loading per square foot, the determination of this factor is of prime importance. When there are a great many individuals required to make such assessments, there is undoubtedly some lack of uniformity and accuracy in this determination. The regulations maintain a "standard" percolation test that is used from time to time particularly where a more complete soil investigation is conducted than is normally possible.

Some states have completed an engineering soils assessment of their soils and have applied this broad knowledge to the extent that they will restrict development to those areas where the soils are classified as being acceptable for subsurface sewage disposal. It is assumed that development of any kind is prevented in areas generally unsuitable but it would be interesting to know how this is accomplished. As previously mentioned, municipal official plans zone the land in various categories; but whether onsite sewage systems are acceptable within the area of the plan cannot be specifically covered before it is approved, and each application for sewage disposal is treated on its merits. In other words, an individual owning land in an area designated as a "hazard" area by the Natural Resources Ministry because of the prevalence of flood plains, steep slopes, marshlands, etc., may obtain approval for a sewage system providing there are no unacceptable conditions in the specific location where he wishes to build. On the other hand, the Ministry of Housing should not even consider a subdivision development in such an area, unless the official plan was amended, and should therefore not circulate the plan for approval.

Because the percolation test is very time-consuming, of doubtful application to the more granular soils and, if our findings are correct, productive of widely varying results in several tests conducted in close proximity, it is not generally used unless to provide an indication of the soil behavior where conditions are judged to be marginally acceptable, or to help in finding the correct solution should there be disagreement between the applicant and the approving authorities. It is our experience that only a few consulting engineering firms retained to prepare an application for development where onsite sewage is proposed are experienced in assessing surface and subsurface conditions for this purpose; some have produced wild claims for the percolative capacity of soils that the approving authorities know from previous experience to be inaccurate.

The assessment of the soil percolation time is done in Ontario by use of field recognition of soil types and strata to a depth of 5 feet, preferably by the excavation of a test pit as opposed to bore holes, supplemented by laboratory tests and field percolation tests, as required.

Water Table

Determination of water table is equally difficult and may indeed be the more important as far as the effective functioning of the leaching bed is concerned. We have yet to find a practical substitute for the method that uses knowledge of the area and a study of the drainage pattern, vegetation and soil strata coloration, in order to modify any actual water table, or absence of water, found at the time of lot inspection.

Maximum Size of Subsurface Disposal Systems

A problem for which we do not have a standard solution, and for which we are adopting solutions based on the merits of each case, is the determination of the maximum size for an individual subsurface disposal system. There are both theoretical and practical considerations involved— for example, whether the size should be based on not exceeding a specified daily sewage flow, or whether it should be limited to the capacity of a standard leaching bed considered to be of the maximum size and layout from the standpoints of ensuring even distribution of effluent and simplicity of construction. A bed of standard "maximum" size would have a varying daily sewage capacity depending on the soils in and below the bed. Other considerations concerning larger subsurface disposal systems relate to:

- The limiting practical size of a septic tank
- The volume of the dose in relation to the internal volume of the distribution pipes
- The length of time within which a dose must be delivered to a bed to ensure as even a distribution as possible
- The time interval between doses or, alternatively, the maximum number of doses per day
- Under what circumstances a standby area, of similar size to the bed area, should be set aside as a spare or insurance in the event of bed failure, and whether the spare bed should or should not be constructed initially.

Equipment Acceptance

A continuing problem is the unavoidable involvement with the public, approving agencies and manufacturers regarding the acceptability of new proprietary systems. The Ministry's responsibilities are essentially the protection of the environment and public health. Prefabricated sewage systems or components are examined with this in mind, assuming they are operated in accordance with the manufacturers' instructions. There is, however, a continual demand for comments of the "Consumer's Report" type to answer the question "Is the system a good one," and the more pertinent one, "Which systems does the Ministry recommend." This is a tricky subject for government representatives to answer, and we endeavor to avoid the recommendation of specific models for obvious reasons. Testing of components, such as electrical components, is arranged by the manufacturer through the Canadian Standards Association, and acceptability from a fire safety viewpoint is cleared by Underwriters Laboratories. However, performance cannot be tested unless standards are written, and this would require, for example, the preparation of standards covering the performance of each of the various forms of toilet.*

A more recent development involves systems that would treat sewage from individual residences or from apartments and condominiums to the point where the quality of the effluent was acceptable either for reuse for a variety of purposes including lawn watering, toilet flushing and bathing, or for "dry ditch" or surface discharge. Indeed the federal government with the Ontario Research Foundation has been working on such a system for some years and, more recently, a major real-estate developer had a system designed and is testing it in an apartment building. At the present time surface discharge of sewage is not permitted and there are no water quality standards for such disposal.

RESEARCH

Our Applied Sciences Section continues to pursue answers to the problems of onsite sewage. Some of their current activities include:

1. Final testing of filter beds for the treatment of septic effluent. Various types of readily available sands have been tested for almost 7 years. The beds are now being tested to destruction because we must give up the property. The beds were underdrained to permit analysis of the effluent. In addition to the sand, quantities of red mud mixed with a sand layer have been tested for the purpose of removing phosphates. Red mud is a waste product from the aluminum smelter at Arvida and contains oxides of calcium, aluminum and iron. A report on performance is available (Publication No. 53, December 1974). Subsequently, the use of limestone layers to remove phosphates was tried but is not included in the referenced report.

2. A test area to assess the relative rate of movement through the soil of different types of effluent has been in operation for 3 years. It involves three separate buried distribution pipes, one fed with septic effluent, one with effluent from a waste stabilization pond, and one from

---

*Note: National Sanitation Foundation Standards are used to evaluate performance of wastewater treatment equipment in the U.S.: NSF Standard No. 40 relating to individual home aerobic wastewater treatment plants; NSF Basic Criteria C-9 relating to special devices used for treating wastewater; NSF Standard No. 24 relating to plumbing system components for mobile homes and recreational vehicles; and NSF Standard No. 23 relating to marine sanitation devices. An NSF standard for wastewater recycle and water conservation systems is expected to be adopted in early 1977. This standard will cover blackwater systems with water and nonwater flushing media, composting toilets, and greywater treatment devices.

an aerobic treatment plant. The test is for movement through the soil rather than clogging. A report is not yet available.

3. At another location a bed of the raised type, designed in accordance with our regulations, is in operation and being monitored as to performance. It is underdrained; thus, the evapotranspiration can be assessed, although the design was not specifically for this purpose.

4. A large bed of 10,000 lineal feet of distribution pipe has recently been completed to study various problems related to large systems.

5. Full-scale sand filters are being tested to determine permissible loading rates on various soils.

6. Studies of the movement of contaminants through the soil have been performed in previous years, with emphasis on the movement of phosphates using radioactive tracers (reports are available). As part of the study of the pollution of the Great Lakes from land use activities for the International Joint Commission, similar studies have been conducted and are in process. For this purpose, a census of the number and location of onsite sewage disposal systems was made and a report prepared.

7. A report is in the preparation stage on a test conducted to assess the feasibility of adding chemicals (aluminum sulphate) to the sewage entering the septic tank, so as to precipitate the phosphates in the tank.

8. A number of systems located in clay soils have been selected for monitoring and will be studied over a period of time.

10

THE NEED FOR IMPROVING SEPTIC SYSTEM REPAIR PRACTICES

William L. Mellen

    District Supervisor
    Lake County Health Department
    Waukegan, Illinois 60085

Recently a man in Lake County, Illinois, spent about $3,500 attempting to repair the septic system for a new single-family home on his estate. Four contractors and three Health Department inspectors later, all he had to show for his investment was a Notice of Violation from the Health Department.

Should health departments make recommendations for septic system repairs, or leave it to septic contractors? To whom can the homeowner turn for advice on repairing his failing septic system? Few public health officials will commit themselves beyond reciting and enforcing ordinances to protect the health of the community. Most contractors receive little or no formal training in analysis and repair of failing septic systems. The repair work they do is often guided only by health department rules and the profit motive.

Few public health personnel have had any formal training in the field of septic system installation and repair. Sanitary engineers, for example, are well educated in the design of sewage treatment plants, but most will acknowledge they have had little or no training in the design of individual sewage disposal systems.

Before one can give a professional opinion on how to repair a failing septic system, he must be well trained and investigate the matter thoroughly. As with many other complicated system analyses, a check list is helpful in evaluating the problem. The investigator should learn as much as possible of the history of a particular septic failure before the analysis of the problem begins.

If the house has a basement, the plumbing should be inspected to ensure that footing drain water is not pumped into the septic system. If there is a laundry sump, it must be sealed to prevent infiltration from the footing drains. Water softeners, humidifiers and air conditioners can generate large amounts of clear water and should not discharge into a septic system.

It is important to check the float valves and water levels of all toilets. Sometimes water levels are too high, or a float valve sticks in an open position and excess water flows into the overflow pipe.

If it is determined that the house plumbing is in good order, the septic tank system should be inspected. Footing drains have been found to be connected directly to soil pipes in those situations where there is plumbing in the basement or lower elevations. Footing drains above the sewer line can also add to infiltration. If no outlets for foundation drains are found, a thorough check of the septic tank, after rainy weather, is essential.

Determine whether or not the tank is in a depressional area with a high water table. Although tanks are referred to as being "watertight," inspection ports and lids are seldom cast in one piece, leaving gaps for infiltration. If the septic tank is located in a depressional area, or an area subject to standing water or a high water table, a septic failure can occur, unless the area is artificially drained.

The area around the septic field should be examined for downspouts, sump pumps, drainageways, parking areas, drywells, or depressional areas that are adding water to the seepage area.

The next step is to expose the dropboxes or distribution box. If a seepage line is dry and the system is failing, proper distribution of effluent in the system is obviously not occurring. Deep lines running perpendicular to the contours are disastrous, and the system should be abandoned. It may be dug up, backfilled, and reconstructed on the contour if there is no area other than the original field for expansion.

On sloping ground with the dropbox system, the last line should be the only line where a failure occurs if water is being properly distributed in the system. Some visual signs, such as poor seeding on top of the seepage field, can indicate compaction of the soil—another potential problem.

A soil inspection can be useful if the investigator understands the information the soil will give him. A soil core can be obtained with the use of a soil sampling tube. If only the percolation test is used, it involves a period of 2 to 3 days if the results are to give any true information, and then they only indicate the sizing of the field.[1] The soil sampling tube is used to remove a 4-foot-long core from two or three locations in the area of the septic field. The core is then examined for color, structure and texture at various depths. The core is laid on the ground in the same order as taken, in order to measure the depth of each horizon as determined by changes in characteristics. It is possible to determine the texture of the soil at the proposed location of the seepage lines, and 6 to 12 inches below the lines. Normally, a shallow system, no deeper than 24 inches, is recommended; but it is necessary to determine the color and texture at a greater depth to classify the soil.

Soil texture can be determined by a method recommended for vocational agriculture students in the Illinois Publication VAS 4052.[2] Moisten a sample of soil to a consistency of working putty. From this sample, make a ball about ½ in. in diameter. Holding the ball between the thumb and forefinger, gently press the thumb forward, forming the soil into a ribbon. If the ribbon forms easily and remains long and flexible, the sample is classified as a fine texture soil (clay or clay loam). If the ribbon forms but breaks into pieces ¾ inch to 1 inch long, it is classified as a moderately fine texture soil. If the ribbon is not formed and the sample breaks into pieces less than ¾ inches long, it is probably a medium textured soil, as a silt loam. If a ribbon is not formed and the soil feels gritty, it is a moderately coarse texture. If the sample consists almost entirely of gritty material and leaves little or no stain on the hand, it is considered coarse texture. A small amount of water is all that is needed to perform this fairly accurate test.

Internal drainage and aeration of a seepage area must be evaluated. The degree of internal drainage and aeration is dependent upon the depth of the seasonal and permanent water tables. The color of the subsoil provides an important clue to the internal drainage and aeration characteristics of a soil. A well drained soil has brown or yellowish-brown color subsoil, and no gray mottling. Mottling is a mixture or variation of soil color. In soils with restricted internal drainage, grey, yellow, red and brown colors are intermingled giving a multicolored effect.[3] The water table fluctuates below the subsurface nearly 12 months of the year. Moderately well drained soils have an upper subsoil above 30 inches that is brown or yellowish-brown and is free from grey mottling. Grey mottling or color may occur below 30 inches. If the water table is as high as the lower subsoil during wet seasons of the year, some type of subsurface drainage will be needed. Curtain drains are used to intercept downslope movement of subsurface water in this type of soil.

Somewhat poorly drained soils have grey and brown-grey mottled subsoils. The wet season water table includes all of the subsoil, and subsurface drainage is definitely needed. Poorly drained and somewhat poorly drained areas have a grey or olive-grey color subsoil. This subsoil is a dull color but may include a few brown colored mottles. The wet season water table is at or near the soil surface. Without improving drainage, these soils are too wet most of the year for septic systems.

The investigator should be familiar with at least three of the eight classes of structure. By examining the top 12 inches and breaking the soil, one can determine whether compaction has taken place. Compacted soil is hard to break and has a massive structure; compaction has destroyed the original structure and has stopped the movement of water.

If a platey structure, which is similar to a stack of dishes, is found near or just below the area of the seepage lines, the soil profile has a clay pan. This clay pan can prevent the downward movement of water, even though the profile has a medium texture topsoil.

Blocky, prismatic and columnar structures, if maintained under wet conditions, have little effect on water movement and are satisfactory for seepage areas. Topsoil color also helps to determine drainage classifications. Dark black soils are usually poorly drained.

Surface and subsurface runoff should be examined in areas of all soils. Soils that seem to be moderately well drained can have the surface water movement restricted by driveways, roads or other structures built upon the land. This must be taken into consideration when reviewing the problem of failing septic systems.

Using Table I, the septic system contractor can now make recommendations on field size and other features that may be necessary to ensure a properly operating system.

The location of the repairs to existing systems is sometimes limited by lot size. If at all possible, a new area should be used. If the original system was without failure for 10 to 20 years, a diversion box should be installed between the tank and the field, and a new field added in a completely separate area. This will allow the resting of the old septic system. Recent studies have shown that an old septic system will rejuvenate itself if allowed to rest.[4] I have found that if a septic system is rested two years, it will work satisfactorily when put back in operation.

Plugged systems are being rejuvenated with hydrogen peroxide in experiments by the Lake County Health Department in cooperation with Dr. John Harkin of the University of Wisconsin.[5]

There is also some evidence that systems that have been damaged by working the soils when they are wet can be rejuvenated by resting.[4] I am sure this is related to the depth of compaction by the equipment and the amount of frost and roots in the area. Some soils, however, if severely compacted during wet weather construction, cannot be rejuvenated even by nature.

If a system is too deep in the ground and installed in a clay subsoil, a lift station can be used. The gravel and the dropbox system must be raised to within 6 inches of the ground surface. A simple lift station consists of an all cast iron or other approved submersible pump in a small

INDIVIDUAL ONSITE WASTEWATER SYSTEMS 91

Table I. Identification of Soil for Individual Sewage System Design

	Coarse Texture Silt loam over sand or gravel	Moderately Fine over Coarse Texture Silt & silty clay 0-36 in. over coarse texture sand or gravel	Moderately Coarse Texture Silty material over sandy loam or gravel loam	Medium Texture Silty material over silt	Moderately Fine Texture Silt material or silty clay; blocky structure; 30 in. or more	Fine Texture Silt material or silty clay; blocky structure; less than 30 in.
Well Drained Soils Soils with a bright and uniform color subsoil on slopes of 2 to 25%	170 sq ft per bedroom	170 sq ft per bedroom; deep, narrow trenches	250 sq ft per bedroom; 24-30 in.-deep trenches	330 sq ft per bedroom; 24 in.-deep trenches		
Moderately Well Drained Soils Soils with uniform color in top 30 in.; mottled subsoil 30-48 in; on slopes of 2 to 25%	200 sq ft per bedroom; subsurface drainage needed	250 sq ft per bedroom; deep, narrow trenches; subsurface drainage needed	300 sq ft per bedroom; 24-in. deep trenches; subsurface drainage needed	330 sq ft per bedroom; 24-in. deep trenches; subsurface drainage needed	400 sq ft per bedroom; curtain drain	Fill or mounded system; curtain drain
Somewhat Poorly Drained Soils Soils with mottling at 6 to 30 in.; black surface; brownish-grey mottled with yellow and greys; 0 to 5% slope; drainageways	Mounded or fill systems	Mounded or fill systems	300 sq ft per bedroom; 24-in. deep trenches; subsurface drainage needed	400 sq ft per bedroom; subsurface drainage needed or mounded or fill system	440 sq ft per bedroom; curtain drain or mounded or fill system	Fill or mounded system; curtain drain & subsurface drainage
Poorly & Very Poorly Drained Soils Dark colored surface; greyish or white colored subsoil; level or depressed landscape; subject to flooding or ponding. All organic and alluvial soils		Filling or drainage may interfere with laws of natural drainage. Use information for Somewhat Poorly Drained Soils, if allowed.				

septic tank or watertight lift chamber. It is sometimes possible to raise the plumbing in the house and brick up the septic tank walls to raise a deep system. The baffle in the septic tank must also be raised.

When a shallow system is found to have too much pitch in the lines, loam fill is required to raise the grade to the same elevation as the highest point of each line. The topsoil can be removed from above the trenches and the top of the gravel leveled. It is not necessary to replace or realign the seepage tile as water will seek its own level.

If space is available, and a replacement system is to be built later, I recommend that the first system be installed in the lower portion of the lot, so the second system can be installed on a higher elevation. This will ensure that the sludge from the first system does not immediately enter the second system, decreasing its life expectancy.

When repairing a system, pick up only the excess water; do not drain the system into the new one. A high line or dropbox may be used so the old system is being put to 100% use before any water flows into the new seepage lines.

In addition to the need to improve repair practices, there is also a definite need to educate the septic contractor and the environmentalist. I do not feel the *Manual of Septic Tank Practice* needs to be replaced as stated last year by Mr. Joseph Salvato.[6] I feel the *Manual* should be revised to include recent research along with an explanation as to why each requirement is so important. This, with an emphasis on education in undergraduate and graduate public health schools, will lead to septic systems being an acceptable type of sewage disposal.

REFERENCES

1. U.S. Department of Health, Education and Welfare. *Manual of Septic Tank Practice,* Publication No. 72-10020 (1972), p. 4.
2. Oschwald, W. R. and R. L. Courson. "Understanding Soils," University of Illinois Publication VAS 4052, pp. 12-14.
3. Parker, D. "Soil Tester Manual," Wisconsin Department of Health, Division of Environmental Health (1974), Section 3, p. 23.
4. Winneberger, T. J. *The Principle of Alternation of Subsurface Wastewater Disposal Fields,* Vol. V (Hancor: 1976).
5. Harkin, J. M. "Causes and Remedy of Failure of Septic Tank Seepage Systems," University of Wisconsin Small-Scale Waste Management Project, *Second National Conference on Individual Onsite Wastewater Systems,* National Sanitation Foundation (1975), pp. 119-124.
6. Salvato, J. A., Jr. "Problems and Solution of Onlot Sewage Disposal," *Second National Conference on Individual Onsite Wastewater Systems,* National Sanitation Foundation (1975), pp. 39-45.

11

FIELD APPLICATION: SAND MOUND AND EVAPOTRANSPIRATION SYSTEMS

Glenn E. Maurer

>Chief, Division of Sewage Facilities Act Administration
>Department of Environmental Resources
>Harrisburg, Pennsylvania 17120

INTRODUCTION

Individual onsite sewage disposal methods and techniques are rapidly becoming a more recognized long-term alternative to sewage collection and treatment systems. In fact, in many situations the onsite system may be the only cost-effective alternative in rural or semirural areas, remote or isolated villages, recreational facilities, and roadside rest areas. The concept of providing sewer interceptors to resolve *all* existing sewage problem areas, or to provide needed facilities for new growth, is no longer acceptable from a social, economic or political point of view. A greater dependency on onsite methods of sewage disposal in the future goes without question.

The fact remains, however, that only 32% (an approximate figure) of the nation has suitable soils for conventional individual onsite sewage disposal systems. The distribution of the good soil, unfortunately, is not adequate for promoting the unlimited use of onsite systems in rural or semirural areas. As with most physical phenomena, the good and the bad have been mixed.

A need, therefore, exists to explore new or alternative approaches to individual onsite sewage disposal systems. The sand mound and the evapotranspiration (ET) systems are two such approaches that have been researched and are now being applied by some regulatory agencies. Pennsylvania's Department of Environmental Resources is one such agency that has permitted the application of these two approaches to individual onsite sewage disposal. Evapotranspiration, however, was allowed strictly on an experimental basis.

CONCEPTS

The "sand mound" system[1] is basically an above-ground system, utilizing sand media to assist in the renovation of the sewage effluent. The system is designed to take full advantage of the upper soil horizons to supplement the renovation processes which occur in the sand mound. It is important to note that the upper soil horizons (say the upper 24 in.) are the most biologically and chemically active areas of a typical soil. The mound system is dependent on lateral movement and dispersion of the wastewater, especially in soils displaying slow permeability. Lateral movement, or lateral permeability, in the upper soil layers promotes the uptake of nutrients and water by plants during the growing season. Wide dispersal of sewage effluent is important to assist in the dilution of pollutants (such as nitrates and chlorides) for which the soil has limited assimilation capabilities.

The evapotranspiration system is a shallow system employing the concepts of evaporation of effluent to the atmosphere and transpiration of effluent by plants. It is impossible to determine the true amount of evapotranspiration versus infiltration of wastewater into the subsurface when an installation employs a combination of these two concepts. Therefore, for the purposes of this paper, I will confine my comments to "total evapotranspiration of wastewater"[2] utilizing an impermeable liner in tight soils or soils having a seasonal high water table.

SOIL SUITABILITY

Subsurface disposal of sewage depends greatly on the soil mantle to filter and renovate wastewater. Soils are used to disperse, filter, assimilate and biodegrade contaminants (either biological or elemental) which are found in wastewater. The soil must react to these compounds to the extent that pollution or environmental health hazards will not occur.

It must be emphasized that many soils are either excessively wet (Figure 1), shallow to bedrock, severely sloping or too rapidly permeable to adequately renovate the wastewater. Under such circumstances, even alternate subsurface systems such as the elevated sand mound will not be acceptable methods for sewage disposal.

In Pennsylvania, a regulatory prerequisite exists which requires a 20-inch depth of suitable soil on an areal basis for a site to qualify for an onsite

Figure 1. Excessively wet soil; unsuitable for onsite disposal of sewage.

sewage disposal system. This, in effect, excludes floodplain soils with a high flooding hazard and somewhat poorly, poorly, and very poorly drained soils on upland sites. Addition of slope limitations to the unsuitable soil categories removes approximately 50% of the land area of Pennsylvania for the utilization of onsite sewage disposal systems. If it were not for alternate systems, such as the sand mound, the percentage of unsuitable soil would rise to approximately 67 to 70% statewide.

The evapotranspiration system employing an impermeable liner precludes soil limitations, to an extent, except under the most severe circumstances. Of course, when you exclude soil infiltration from evapotranspiration, climatic factors become much more critical.

THE ELEVATED SAND MOUND

The concept of the sand mound was first introduced into Pennsylvania in the early 1970s under a policy statement which imposed certain criteria for its use. As a result of this policy, several hundred mounds were installed in Erie County and the Pocono region of the state. In 1972, a Technical Advisory Committee to the Department of Environmental

Resources was appointed to study alternate methods of sewage disposal. This committee evaluated the early mounds installed under the policy statement, as well as the studies performed on sand mounds at the University of Wisconsin under the Small-Scale Waste Management Project. The Technical Advisory Committee concluded that the sand mound was environmentally sound, and its use as an onsite sewage disposal technique should be expanded in Pennsylvania. As a result, revisions to the Rules and Regulations[3] were made in September 1974 to incorporate construction standards (Figure 2) for the sand mound system.

Figure 2. Construction of sand mound; sand being placed in mound.

To date, approximately 2,500 sand mound systems have been installed and are in use in the state. The majority of the systems are for individual residences; however, a number are being utilized for smaller institutions and commercial establishments. A sewage disposal system incorporating a series of sand mounds is currently in the design stage for a small subdivision of 22 lots. This system will involve a community collection system and will be closely monitored over a period of years.

The utilization of sand mound systems in Pennsylvania has not been without problems. There have been a number of confirmed malfunctions since the establishment of the construction standards in 1974. In every case, however, the malfunction was traced to improper enforcement, in that a system was permitted which did not conform to the design standards, or problems associated with faulty installation or improper maintenance of the system.

Regardless of the "track record" of the mound system in Pennsylvania, its use has been the center of considerable controversy. Issues of aesthetics, costs, availability of sand media, growth—no growth, complexities of installation and maintenance, and the soundness of the system have surfaced.

• Individual aesthetic values have received considerable attention with the advent of the mound system in Pennsylvania. People seem repulsed to the notion of a mound in the lawn which represents their sewage disposal system. Rumor even has it that an uninformed individual attempted to remove that pile of topsoil, that the builders so nicely left behind, only to find it was his sewage system. Complaints or concern regarding the aesthetics of mounds are generally without foundation. Mounds can be planned and designed to blend (Figure 3) into the lot, or to actually complement the setting of the home. Some difficulty can be encountered, however, in the "blending in" of a mound for a repair to a malfunctioning system, or for use on a small level lot.

Figure 3. Sand mound blended into surrounding lot.

• Costs have also been the subject of discussion with regard to the mound system. Prior to the establishment of construction standards for mounds, it was common knowledge that this system would be two to three times the cost of conventional methods of sewage disposal. These increased costs can be attributed to complexities of construction and the availability of sand media. Initially, total costs for the mound system for a single-family residence ranged from $1,500 to $5,500. These figures are based on a survey conducted by the Department of Environmental Resources in August 1975. At the $5,500 price, there is no question that some profiteering was taking place. More recent data suggest, however, that the price of mounds in Pennsylvania has leveled to approximately $1,800 to $2,400 for a single-family residence. The reasons for this cost reduction are: more competition between installers, increased availability of sand media, and increased experience in the construction of mounds.

- The availability of sand media has posed some problems with the incorporation of the mound as an alternative sewage disposal system. Pennsylvania requires a "dirty sand" as a media. The amount of "fines" in the media is set by regulation. The fines provide for chemical and biochemical interaction with clay and other dissolved materials. The fines also retard movement of effluent through the mound to a degree, providing additional time for biological action. The availability of sand in the quantities desired initially presented some problems; however, it is now readily accessible in most areas of the state.

- Without a doubt, the advent of the mound system in Pennsylvania has opened more land for development purposes. An interesting dichotomy has resulted. In some areas of the state, the "no growth" issue is quite strong, especially in the southeast. People view the introduction of the mound system as a threat to the remaining open space and as promoting sprawl-type development. Often these limited-growth advocates cry that the system is not sound and ultimately will lead to pollution or environmental health problems. On the other hand, many areas of the state, especially the northwest and northern tier counties, have extremely poor soils. These soils are frequently so poor that even the sand mound does not provide a suitable means of sewage disposal. Thus, the people from these areas of the state want a modification to the standards which would provide for the expanded use of the mound system.

- The complexities of installation and maintenance of the sand mound system are real and must be carefully considered by regulatory agencies. Needless to say, however, any onsite system—conventional or alternate—is subject to failure if improperly designed, incorrectly installed, or not maintained on a regular basis.

EVAPOTRANSPIRATION

Two experimental total evapotranspiration systems have been installed and evaluated by the Department of Environmental Resources with respect to their applicability in Pennsylvania. The first of these systems was installed in June 1974 for a drive-in branch bank; the latter, about one year later for a VFW club. These evapotranspiration systems replaced a holding tank at the bank and a severely malfunctioning onsite system at the VFW. These systems were monitored over the last 2½ years.

The systems utilized an impermeable liner, rendering them totally dependent on evaporation and transpiration of the sewage effluent. The system at the bank was installed under the supervision of Dr. Alfred P. Bernhart from the University of Toronto. The system at the VFW was installed to Dr. Bernhart's specifications, although he was not present at the site during installation. Both systems incorporate a holding tank at the end of the ET beds which collect and hold overflow in the case of system failures.

A problem was encountered with the ET system for the branch bank in that the rim of the liner in one corner of the bed was lower than the liner for the complete bed. Thus, the system would surge (Figure 4) at this low point during heavy rain or excessive effluent discharge to the system. In addition, the ET system at the bank repeatedly discharged effluent diluted with infiltrating rainwater to the holding tank. Such overflows, in effect, constituted failure of the system.

Figure 4. Malfunctioning evapotranspiration system.

The experimental system for the VFW club has been in use for a shorter time duration than the system for the bank. The VFW system consists of two ET beds connected in series. The lower bed receives excess wastewater from the upper bed during peak flow or excessively wet periods. Observations conclude that the lower bed has periodically overflowed (malfunctioned) to the ground surface bypassing the backup holding tank.

The contention that evapotranspiration is an effective means of totally disposing of sewage effluent in a climate similar to Pennsylvania remains unsubstantiated according to the results of these two experimental systems. Additional data supporting the negative conclusion of these experimental systems can be found in the work accomplished at the University of Wisconsin by Tanner and Bouma.[4]

SUMMARY

The sand mound and the evapotranspiration systems are only two alternatives to onsite wastewater disposal which are being considered nationally. Many other methods for onsite disposal of sewage are presently being studied and appear to hold promise for the future. It is important that work continue in the development of such alternatives, keeping in mind that dependency on such systems will be necessary for the future.

The sand mound system appears to be an environmentally sound and cost-effective approach to sewage disposal. The evapotranspiration system does not appear to be suited to the climatic conditions of Pennsylvania; however, it may prove satisfactory to regions of the nation where arid to semiarid conditions prevail. At this time, additional research is being conducted regarding the application of these two systems under varying conditions. The University of Wisconsin and the Pennsylvania State University are presently involved in one such study of the sand mound system. The Virginia Polytechnical Institute is presently researching the evapotranspiration system. Hopefully, the findings of institutions like these will establish additional breakthroughs in the field of onsite wastewater disposal.

The institutional and administrative arrangements for administering a program of onsite sewage disposal continue to be a major obstacle. Regulations and standards for conventional or alternate systems are of little value if they are not understood or not properly enforced. Regulatory programs can utilize a variety of management techniques which are effective and insure proper design, installation and maintenance of onsite sewage disposal systems. Public information and education cannot be overlooked in the development of such a program. All too often, the uninformed public has caused serious problems in relatively sound programs for the protection of our environment.

Regulatory agencies must carefully consider the institutional and administrative arrangements for program implementation. They must understand and consider the ultimate impact of the program on the public. Such realities are essential to the development of meaningful programs for the control of individual onsite wastewater systems.

REFERENCES

1. "On-Site Wastewater Disposal for Homes in Unsewered Areas," Small-Scale Waste Management Project, Publication No. R2533 (September 1973), pp. 13-16.
2. Bernhart, A. P. "Treatment and Disposal of Wastewater from Homes by Soil Infiltration and Evapotranspiration," Second Edition, (1973), pp. 47-51.
3. Pennsylvania Department of Environmental Resources, Rules and Regulations, Chapter 71, "Administration of Sewage Facilities Program," Chapter 73, "Standards for Sewage Disposal Systems," Harrisburg, Pennsylvania (1974).
4. Tanner, C. B. and J. Bouma. "Evapotranspiration as a Means of Domestic Liquid Waste Disposal in Wisconsin," University of Wisconsin, Madison, Wisconsin (February 1975).

12

EFFLUENT QUALITY CONSIDERATIONS AFFECTING THE USE OF SAND FILTERS AND OXIDATION LAGOONS

Michael Hines

    Assistant State Sanitary Engineer
    Illinois Department of Public Health
    Springfield, Illinois 62761

INTRODUCTION

During the last 30 years, large numbers of American people have moved from the rural environment to urban areas. The growth rate of residential centers in these urban areas has been faster than the ability of the municipal governments to provide central wastewater collection and treatment facilities to serve these individuals. As a result, we are seeing more and more large residential areas being served by individual, private sewage disposal systems. In the large urban area of northeastern Illinois, it has been estimated that, during the last several years, one-half of all homes constructed in that area were served by private sewage disposal systems.

As larger numbers of people develop residences in small land areas, the problem of disposal by surface discharge of effluents from private sewage systems becomes critical. Public health authorities have for years been attempting to develop new—or redevelop old—private sewage disposal concepts that would result in effluents which could be discharged to surface water courses without providing a threat to the public health of the community. The use of oxidation lagoons and modifications of the septic tank sand filter system have shown promise in accomplishing this goal.

Water pollution control agencies, which in many states have been completely divorced from public health agencies and, in many cases, public health philosophies, have been very actively promulgating and revising effluent and stream quality standards which would affect the private sewage disposal systems as well as the large municipal and industrial waste disposal systems. We are beginning to see a direct confrontation in many areas between the quality standards of the water pollution control agencies, and the design and operation standards for private sewage disposal systems enforced by many public health agencies.

DEVELOPMENT OF EFFLUENT AND STREAM QUALITY STANDARDS

During the middle and late 1960s the American public became acutely conscious of environmental degradation. As a result of their concern for the continued pollution of our rivers and lakes, they demanded that more stringent controls be exerted over waste discharges. Water pollution control agencies were created where in many cases no such agency had existed in the past. In other areas existing water pollution control agencies were completely reorganized. Unfortunately, many of these agencies were staffed with environmentalists rather than environmental health professionals. Also unfortunately, in many cases administrative decisions regarding effluent quality standards, treatment system design standards, and operation control standards were made by individuals with legal, political or business management backgrounds, rather than by environmental health professionals.

The effluent and stream quality standards developed in many areas of this country have very little relation to protection of the public health. Many of these standards were developed in an atmosphere of hysteria rather than as an attempt to provide the best possible surface water quality with the minimum possible impact upon the economy.

In the state of Illinois exist some of the most stringent effluent quality standards anywhere in the United States. The standards are very complex and contain different parameter values for different areas of the state, and for different dilution ratios provided by receiving streams. It is safe to make the assumption in Illinois that unless a stream is large enough to have a name, it is classified as an intermittent stream and given no credit for providing any dilution ratio to sewage effluents.

Any sewage effluent discharging to such an intermittent stream is required to have an effluent BOD less than or equal to 4 mg/l, effluent suspended solids of less than or equal to 5 mg/l, and fecal coliform levels less than or equal to 400 per 100 ml, as a maximum. These three parameters are the only parameters which have currently been adopted as effluent standards.

In addition to these effluent parameters, however, the quality of the water in the receiving stream must be maintained within certain limits.

The stream standards cover the areas of dissolved oxygen, ammonia, phosphorus and fecal coliform. The dissolved oxygen concentration must never be below 5 mg/l, and must not be below 6 mg/l for any 16-hour period of the day. The ammonia concentration must be less than or equal to 1.5 mg/l, and the fecal coliform levels must not exceed 200 per 100 ml, as a geometric mean. Phosphorus standards exist for all surface water impoundments. The standard for phosphorus is 0.05 mg/l within the reservoir, or within any river at the point at which it joins the reservoir.

It becomes intuitively obvious that a roadside ditch will have as its only flow through most of the year the sewage effluent which may be discharged to it. As a result, the stream standard in that ditch becomes the effluent standard for that ditch. Also, if that dry ditch lies within the drainage basin of a surface water impoundment, the Illinois water pollution control authorities stipulate that any sewage effluent to that dry ditch must meet the 0.05 mg/l phosphorus limitation.

To further confuse the standards picture, we have in Illinois what is called the "Pfeffer" exception. This exception is available to small sewage disposal systems to allow them to discharge effluents with BODs of less than or equal to 10 mg/l and suspended solids of less than or equal to 12 mg/l so long as the previously discussed stream standards and bacterial standards are not violated by such effluents. Additionally, three-stage oxidation lagoon systems serving less than 2500 population equivalents are allowed to discharge effluents with BODs of less than or equal to 37 mg/l. These systems also, however, must meet all the previously defined stream standards and bacterial standards.

Obviously, the frustration facing the designer, owner, operator or public health official charged with regulating private sewage disposal systems becomes considerable.

HISTORICAL USE OF SAND FILTERS AND OXIDATION LAGOONS

Sand filters have been used as sewage treatment systems in Illinois for 50 years or more. Most of the earlier systems were standard septic tank buried sand filter installations, or septic or Imhoff tank intermittent sand filter combinations. These systems were almost always located in rural areas with very low system concentrations. The effluents from these systems were discharged generally to roadside ditches or farm tiles with ultimate discharge to small intermittent streams.

While the effluents from these systems would not meet today's stringent standards, they generally had no serious impact on the receiving streams, did not cause nuisance conditions to develop, and did not result in any significant threat to the public health. In fact, the effluents from these small systems usually soaked into the ground within a few hundred feet of the point of discharge.

Within the last several years, these standard sand filter systems have been modified to recirculating sand filter systems. This type of system has proved to be quite acceptable for use at private residences in highly developed residential areas. Experience has shown that, if these systems are properly designed, installed and operated, they can produce effluents which will meet the stringent effluent and stream quality standards. Experience has also shown, however, that in many cases, they do not receive the operation attention necessary to insure the high effluent quality.

The use of oxidation lagoons for small-scale sewage treatment systems has not been extensive in Illinois. There are several such systems located primarily in southern Illinois which have been evaluated by both state and local health authorities. The oxidation lagoon is a very simple system which does require a relatively large land area. Most of the systems utilize single-cell lagoons which receive either raw domestic sewage or septic tank effluent. These systems operate essentially as nonoverflow systems during the hot and dry periods of the year, and as overflow systems during the remainder of the year.

So long as they are properly designed and well operated, these systems produce a normal oxidation lagoon effluent. These effluents, while they are not of as high quality as some of the sand filter effluents, have been found to be satisfactory for discharge to intermittent streams in isolated areas. The organic material remaining in the effluents is associated primarily with algae, and has not produced any nuisance conditions or any particular public health hazards. These effluents, however, do not meet the stringent effluent quality standards enforced in Illinois.

As with the sand filter systems, many homeowners will not provide the attention to the operation and maintenance of the lagoon systems and they become, in many cases, overgrown with weeds and bottom vegetation and develop into excessive mosquito-breeding habitats.

DISCUSSION

One of the problems that must be addressed is the question of whether or not the effluent quality standards and stream quality standards established by regulatory agencies are reasonable. Standards that would be reasonable to the water quality purist would not necessarily be reasonable to those of us concerned with public and environmental health. The role of economic considerations and provision of additional treatment processes is one that is beginning to receive more attention; however, it does not seem to be intelligent attention. Design engineers are required to do cost-effectiveness analysis of additional treatment processes. They are not allowed, however, to determine whether or not it would be cost-effective to provide the additional treatment. They are required to determine *which* of the approved additional treatment processes would be the *most* cost-effective. It is only a matter of time until the public begins to demand that somebody determine whether or not it is cost-effective to require the tertiary and quaternary treatment processes now being demanded by regulatory agencies.

Effluent BOD standards have been set with the philosophy that within certain limitations, the greater the dilution ratio provided by the receiving stream the higher the allowable effluent BOD limit would be. This would seem to be a reasonable approach and would allow the use of the receiving stream as a part of the treatment process without degrading the quality of the receiving stream below certain acceptable limits. A problem arises, however, in that these effluent standards are not established on a case-by-case, plant-by-plant basis, but for an entire state. As a result, small-scale waste treatment systems discharging to dry ditches are forced to provide effluents of almost pristine quality. In most cases, these effluents disappear within a short distance of the discharge point. The very restrictive BOD and suspended solids standards cannot be reasonably justified for the small plants on a statewide basis.

There has always been—and will continue to be—a controversy regarding the real threat to the public and environmental health posed by discharge of microorganisms from sewage treatment plant effluents. At both the federal and state levels, fecal coliform limits have been established. In order to meet these limits, disinfection of all effluents is required. There are no more than one or two documented cases of waterborne disease outbreaks involving swimmers utilizing waters polluted by sewage organisms. In fact, British researchers who have done considerable work in this area, have essentially concluded that the major threat of waterborne disease outbreak occurs if swimmers ingest a certain amount of the actual fecal material which could be present only if raw or partially treated sewage was being discharged to a water use recreational area.

The discharge of effluents from small-scale sewage treatment systems, if not disinfected and if allowed to accumulate and stand in highly populated areas, could pose a threat to public health as a result of access to the sewage effluents by children and pets. In the rural environment with isolated discharges, however, it is doubtful that these discharges would pose any significant threat to the public health. In any event, the decision on disinfection of small-scale sewage treatment system effluents should be made on a case-by-case basis and should not be made on a statewide basis. Also, the requirement that these or any sewage effluents be disinfected 12 months out of the year would not seem to be reasonable in areas with periods of freezing weather, such as we have in Illinois.

The problem of discharge to the aquatic environment of nutrients from wastewater treatment plants has resulted in controversy within the sanitary engineering profession for years. In any area of the country where farming operations exist, the nutrient load discharged to the aquatic environment from sewage treatment systems is insignificant in comparison with the nutrient load discharged from agricultural operations. Regulatory agencies have discovered that it is extremely difficult, if not impossible, to regulate the discharge of nutrients from agricultural operations. They can, however, regulate the discharge of nutrients from wastewater treatment systems.

The resulting nutrient standards, which have been established in most areas of the country and certainly in Illinois, are such that small-scale sewage treatment systems cannot possibly meet the standards without the addition of very expensive additional treatment devices. These devices provide no additional protection for the public health and do not provide any measurable protection to the aquatic environment because small-scale sewage systems themselves have very little impact upon the aquatic environment.

A second major question with which we are faced is, "Can small-scale sewage treatment systems, particularly the oxidation lagoons and sand filter systems, provide effluents which would meet the effluent quality and stream quality standards which have been established?" While it is not desirable to generalize, it must be generally recognized that the answer to this question is "no."

Without a doubt, there are particular examples of properly designed, constructed and operated lagoons or sand filter systems which do meet the BOD and suspended solids standards of 30 mg/l and 37 mg/l, respectively. Even these unique systems, however, cannot meet the nutrient standards.

Oxidation lagoons and the normal buried sand filter systems will not be able to meet effluent BOD and suspended solids standards of 10 mg/l and 12 mg/l, respectively. With modifications, sand filter systems such as intermittent sand filters and recirculating sand filters could, if properly operated, meet the 10-mg/l BOD and 12-mg/l suspended solids standards rather consistently. Again, however, they could not meet the nutrient standards. None of these systems, with the possible exception of the recirculating sand filter system, can ever hope to meet the 4-mg/l BOD and 5-mg/l suspended solids standards. The recirculating sand filter system with detailed attention to operation can meet these standards. Again, however, it cannot provide an effluent which would meet the nutrient standards.

None of the systems could hope to provide effluents which would meet the bacterial standards unless disinfection were required on each and every system. The major decision faced by those of us involved in the regulation of small-scale sewage treatment systems is whether or not such systems have to meet the effluent and stream quality standards previously discussed. There seems to be a prevailing attitude that there must be some effluent standards established for any waste treatment system. This infers sampling of the effluents, measurement of the parameters for which standards have been established, and the use of the results to determine pass or fail of the treatment system. This, of course, requires access to laboratory facilities, and most local environmental health programs do not have such access. I seriously question the need for actual standards to be set for small-scale sewage plant effluents. It would seem more appropriate to establish exact design, construction and operational criteria which can be easily regulated by environmental health authorities. So long as systems were in conformance with these criteria, the effluents should be satisfactory in terms of effect upon the public and environmental health.

If effluent standards are going to be demanded, I do not think that we should utilize the very stringent effluent standards which have been promulgated by the water pollution control authorities. Many years of sanitary engineering experience have shown us that a small-scale sewage treatment system utilizing secondary treatment can discharge an effluent which, so long as it is within the range of BOD and suspended solids of 30 mg/l and 35 mg/l, respectively, does not cause nuisance conditions in the stream and does not result in any significant public health hazard. The key to this philosophy is the secondary treatment requirement. A septic tank sand filter installation provides secondary biological treatment, both in the septic tank and in the sand bed itself. The oxidation lagoon provides secondary treatment with conversion of the incoming BOD and suspended solids into a form associated primarily with the algae mass, and not with the sewage solids that enter the lagoon.

To establish an effluent quality standard of 30 mg/l of BOD and 35 mg/l of suspended solids, and to require that small-scale sewage treatment systems be designed, constructed and operated in accordance with sound sanitary engineering principles would result in the allowed use of such systems and would not result in effluents which would seriously threaten the health of the public, or damage the aquatic environment. It would still be necessary in local areas, however, to require more stringent treatment and more stringent effluent standards in order to satisfy local problems. For example, in a subdivision with 100 individual residences, each to be served by a small-scale sewage treatment system, it might be desirable to establish a requirement that all systems must utilize sand filtration with effluent disinfection to insure the best quality effluent because the system density would be high.

It is always much easier to establish standards on a statewide or nationwide basis. Regulatory agencies tend to prefer this because it makes their job of regulating individual systems much easier. A much more desirable philosophy is to establish general guidelines on a case-by-case basis. This, of course, requires field evaluation of problem sites and determination of the best guidelines and standards to be developed for those individual sites. This philosophy would result in the maximum protection of the public and environmental health with the minimum waste of natural and economic resources.

SUMMARY

There are many oxidation lagoon and sand filter systems which can meet a reasonable effluent quality standard. There are few if any such systems, however, which can meet the very stringent effluent quality standards currently in existence. None of the systems can meet the nutrient standards which have been established as stream standards in many areas. The very extensive addition of treatment devices required to obtain an effluent from a sand filter or lagoon system which would meet the very stringent effluent and stream quality standards cannot be justified on the basis of economics or protection of public and environmental health.

Decisions affecting effluent quality requirements for small-scale sewage treatment systems should be completely divorced from similar decisions regarding large-scale waste treatment systems. Many small systems are isolated in the rural environment and discharge effluents to surface water courses in areas of very low population density. In other areas, small-scale sewage treatment systems are being provided in high density in areas with corresponding high population densities. In order to provide maximum public and environmental health protection, different effluent quality standards are necessary in each area.

Extremely stringent effluent quality standards established on a statewide or large areawide basis would greatly increase the cost of such systems and cannot be justified from the standpoint of protection of public or environmental health. More desirably, a set of effluent quality standards could be adopted on a statewide or large areawide basis which would insure that the minimum public and environmental health protection necessary was achieved. In areas of high population density or high small-scale sewage treatment system density, more stringent local standards should be authorized and developed. Decisions regarding such standards must be made by environmental health and sanitary engineering professionals, not strictly by legal or political interests.

We have been charged with providing the maximum public and environmental health protection possible with a minimum economic impact. The environmental health profession must redirect its efforts toward accomplishing that goal.

13

TREATMENT SYSTEMS REQUIRED FOR SURFACE DISCHARGE OF ONSITE WASTEWATER

David K. Sauer

>Sanitary Engineer
>Small-Scale Waste Management Project
>Department of Civil and Environmental Engineering
>University of Wisconsin
>Madison, Wisconsin 53706

INTRODUCTION

Many onsite wastewater disposal systems for homes in the United States are constructed on soil types that have contributed to the failure of the conventional septic tank-soil absorption system. Many other homes cannot be built because of poor soil conditions for waste disposal. Home construction where good soil conditions exist also leads to the depletion of prime agricultural land. Surface discharge is an alternative disposal method. However, to avoid environmental and health problems, discharge of onsite wastewater to the surface may require extensive treatment and disinfection.

There are numerous alternative schemes which may be considered by the homeowner to achieve surface water discharge requirements. To make a proper decision on these alternatives, the homeowner must examine both the in-house processes,[1] as well as treatment options which might best meet the local water quality objectives. A general matrix depicting these choices might be similar to that shown in Table I.

Many of the options listed have been examined or are being evaluated by the Small-Scale Waste Management Project (SSWMP). Among those treatment options considered most effective is the intermittent sand filtration of either septic tank or aerobically treated effluent, followed by

Table I. Suggested Processes for Onsite Surface Water Treatment Systems

In-House Processing	Treatment Processes	Water Quality Objectives
Water conservation fixtures	Septic tank systems	BOD
Waste segregation–grey/black	Aerobic processes	Suspended solids
Household product selection	Intermittent sand filtration	Pathogens
Appliance selection	Activated carbon filtration	Phosphorus
Waste/water recycle-reuse	Wastewater lagoons	Nitrogen species
	Clinoptiolite ion exchange	
	Nitrification	
	Denitrification	
	Chemical precipitation	
	Chlorination, iodination	
	UV irradiation, bromination	

disinfection. The major emphasis of this paper will be to examine the effectiveness of this alternative to meet a water quality objective concerned primarily with low BOD, suspended solids, and fecal coliform concentrations.

DESCRIPTION OF EXPERIMENTAL STUDIES

Laboratory and field investigations of the intermittent sand filtration process following septic tank or aerobic treatment processes have been underway since 1973. In addition, disinfection studies in the laboratory and field have been undertaken.

Laboratory Experiments

The objective of the laboratory study was to determine the effect of applied wastewater quality, media size and hydraulic loading rate on sand filter effluent quality, length of filter run, filter clogging and crust development.[2] Tensiometric measurements were also taken to characterize—and possibly predict—filter run lengths. Various maintenance techniques, such as resting, raking or removing the crusted sand surface were also studied as methods of rejuvenating clogged sands.

The laboratory experiments were performed at the Park Street Laboratory and involved the application of septic tank effluent and aerobic unit effluent to 24 4-inch (10.2-cm) sand columns. This study has been conducted since June 1975. Three types of sand, two hydraulic loading rates, and two effluent types have been employed in the study. The effective size and uniformity coefficient of the sands are listed in Table II. The

Table II. Sands Used for Laboratory Experiments

Effective Size (mm)	Uniformity Coefficient
0.19-0.22	3.3-4.0
0.43-0.45	3.0-3.3
0.65	1.4

hydraulic loading rates used in the study were 5 gal/day/sq ft (0.2 m/day) and 10 gal/day/sq ft (0.4 m/day). The wastewater applied to the columns was a simulated household waste treated by a septic tank or an aerobic treatment unit.[3] Wastewater effluents from these units were monitored, and selected parameters are shown in Table III. Wastewater flow rates through the units were designed to follow the characteristic flow pattern of single households.

Table III. Wastewater Characteristics at the Park Street Laboratory

	BOD_5 (mg/l) Mean	TSS (mg/l) Mean
1000-gallon septic tank effluent	53	25-54
Extended aeration aerobic treatment unit effluent	20-34	50-85

The sand columns contained 30 inches (76.2 cm) of sand underlain by 6 inches (15.2 cm) of pea gravel and 6 inches (15.2 cm) of coarse gravel. A freeboard space of 18 inches (45.7 cm) existed above the sand to allow intermittent ponding of wastewater above the sand. The columns were dosed on the average six times per day.

Field Experiments

The objective of the field study was to determine whether household wastewater applied at hydraulic loading rates of 2 to 10 gal/day/sq ft (0.08 to 0.4 m/day) could be adequately treated by sand filters, and to determine the length of filter runs. Various types of maintenance, such as raking the sand surface, removing the top layer of sand, or resting the filter bed for some period of time were also investigated. Disinfection of the sand filter effluents via dry-feed chlorination and ultraviolet light irradiation was also investigated in the field experiments.

The field experiments have been conducted since September 1973.[4] Systems were constructed at three rural homes located on University of Wisconsin experimental farms. The system at the Ashland Experimental Farm employed the sand filtration of septic tank effluent, while the system at the Electric Research Farm involved the sand filtration of an aerobic unit effluent. Disinfection of the sand filter effluent at both of these sites used gravity flow through dry-feed chlorinators. The system at the Arlington Experimental Farm included the sand filtration of both septic tank effluent and of aerobic unit effluent. Disinfection of the sand filter effluents was performed by an ultraviolet light irradiation unit.

All field sand filters ranged from 14 to 16 sq ft (1.3 to 1.5 m^2) in area and contained 24 to 30 inches (61.0 to 76.2 cm) of washed sand. The sand was underlain by 6 inches (15.2 cm) of pea gravel and 6 inches (15.2 cm) of coarse gravel. The effective size and uniformity coefficient of the sand used at each site are listed in Table IV. It is important to note that the sands used at the Ashland Experimental Farm and the Electric Research Farm were also tested in the laboratory column studies. The sand was washed pit run sand which was locally available and relatively inexpensive.

Table IV. Sands Used for Field Experiments

	Effective Size (mm)	Uniformity Coefficient
Ashland Experimental Farm	0.43-0.45	3.0-3.3
Electric Research Farm	0.19-0.22	3.3-4.0
Arlington Experimental Farm	0.28	2.8

The sand filters were enclosed in concrete block basins and were placed below ground level to prevent freezing problems. The top 4 inches (10.2 cm) of the basins were above ground level to allow an insulated and removable cover to be fastened to the top of the filter. The covers prevented the accumulation of debris on the sand surface, reduced odor, eliminated freezing problems, and allowed easy access to the sand surface. An open space of approximately 16 inches (40.7 cm) existed above the sand surface to allow intermittent ponding of wastewater above the sand.

The distribution system for the sand filters consisted of a 2-inch (5.1-cm) plastic pipe with an up-turned elbow located in the center of the bed. A splash plate was placed underneath the outlet elbow to reduce erosion of the sand surface. The collection system at the filter bottom consisted of a 4-inch (10.2-cm) perforated pipe which was vented above the sand surface. Further description of the sand filters can be found in the literature.[5]

The hydraulic loading rates employed for the sand filter studies ranged from 2 to 40 gal/day/sq ft (0.08 to 1.6 m/day), with the rates primarily between 2 to 10 gal/day/sq ft (0.08 to 0.4 m/day). Excessive groundwater infiltration sometimes caused the rates to become exceedingly high. The sand filters were dosed (4 to 13 times/day) by a submersible pump with a controlled volume of wastewater. This was dependent upon the amount of wastewater generated each day.

RESULTS AND DESIGN RECOMMENDATIONS

Septic Tank-Intermittent Sand Filter

Sand Clogging and Maintenance

Throughout the field and laboratory studies, a record was maintained of the infiltration rate of sands loaded with septic tank effluent. Because there were both field and laboratory studies, many curves were generated representing the decline of infiltration rate over a given time period. A detailed discussion concerning this subject will be presented in a forthcoming publication. In this paper, a general discussion highlighting significant findings will be presented.

Figure 1 shows the infiltration rate decline of a sand loaded with septic tank effluent. The data in this figure were generated at the Ashland field sand filter. The sand had an effective size of 0.43 mm and an initial saturated hydraulic conductivity of 565 gal/day/sq ft (23 m/day). The average hydraulic loading rate was 4.85 gal/day/sq ft (0.2 m/day). This figure is shown because it represents the characteristic infiltration rate reduction that was found in both the field and laboratory studies.

The length of this filter run was approximately 150 days; however, it was found from other filter runs and the laboratory sand columns that run lengths varied depending upon the loading rate, the strength of the wastewater, maintenance performed to the sand surface, and the sand size. An important finding was that the pattern of infiltration rate reduction was similar in all of the studies. As shown in Figure 1, a logarithmic decline in infiltration rate occurred leading to continuous ponding of the sand and an ultimate infiltration rate between 0.5 and 1.0 gal/day/sq ft (0.02 and 0.04 m/day).

An initial time period at the startup of the sand filters did not show this logarithmic infiltration rate decline. The length of this time period

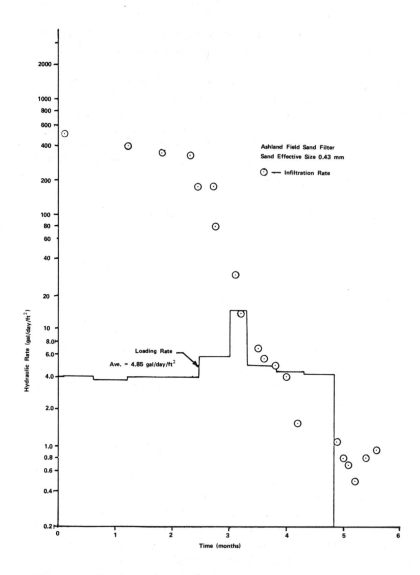

Figure 1. Infiltration rate decline of sand loaded with septic tank effluent.

largely influenced the length of the filter runs. Larger sand size, lower hydraulic loading rates, and lower wastewater organic strengths increased filter run lengths.

Although the true mechanisms involving the purification of septic tank effluent using sand filters have not been delineated, it is speculated that, during the initial startup period, a large increase occurs in the microbial activity within the sand. Products of microbial growth, such as cells, slimes and polysaccharides, accumulate throughout the sand bed. This growth is especially high in the top 2 to 4 inches (5.1 to 10.2 cm) of sand. Eventually these growth products reduce the volume of the sand pores, resulting in the logarithmic decline in the infiltration rate.

The ultimate infiltration rates of 0.5 to 1.0 gal/day/sq ft (0.02 to 0.04 m/day) were found in nearly all the experimental studies. To regenerate the clogged sand beds, various maintenance techniques were tested. Results showed that a physical breakup of the crust which forms in the top 2 to 4 inches of sand along with a resting period was required to adequately restore the infiltration capacity of the sand. The breakup of the crust was performed by raking the top 2 to 4 inches of sand, or by replacement of the crust with clean sand. The resting period was essential to allow the lower portion of the sand bed to aerate and regenerate.

Based on both laboratory and field results with septic tank-sand filter systems, a hydraulic loading rate of 5 gal/day/sq ft (0.2 m/day) is recommended for determination of the required surface area of the sand filter. As a result of the clogging effect of septic tank effluent, it is recommended that an additional sand filter of equal size be installed. Application of effluent onto the sand filters would alternate between the two beds. Time periods of loading and resting have been found to be dependent upon the effective size of the sand. These values are listed in Table V.

Table V. Septic Tank-Sand Filter Operation Schedule

Sand		Loading and Resting Period
Effective Size (mm)	Uniformity Coefficient	(months)
0.2	3-4	1
0.4	3	3
0.6	1.4	5

For example, if a sand with an effective size of 0.4 mm and uniformity coefficient $\simeq 3$ is used, the operation and maintenance schedule is as follows: The entire wastewater load is applied to the first sand filter for 3 months. The flow is then switched to the second sand filter, and the first filter is raked to a depth of 2 to 4 inches (5.1 to 10.2 cm).

Wastewater is reapplied to the first filter after 3 months of rest. After the second 3-month loading of the filter, the top 4 inches (10.2 cm) of the sand is replaced with clean sand. The first filter then rests for 3 months while the second filter is in operation. Total maintenance to each filter involves raking the sand surface and removing the top sand crust once each year.

Effluent Quality

A detailed analysis of the effluent quality from field sand filters loaded with septic tank effluent at an average rate of 5 gal/day/sq ft (0.2 m/day) has been published previously.[4] More recent data have been similar, so only selected parameters are presented in Table VI.

Table VI. Septic Tank-Sand Filter Effluent Quality Data

	Septic Tank Effluent	Sand Filter Effluent	Chlorinated Effluent
BOD_5 (mg/l)	123	9	3
TSS (mg/l)	48	6-9	6
Ammonia-N (mg/l)	19.2	0.8-1.1	1.6
Nitrate-N (mg/l)	0.3	19.6-20.4	18.9
Orthophosphate (mg/l)	8.7	6.7-7.1	7.9
Fecal coliforms (#/100 ml)	5.9×10^5	$(0.5\text{-}0.8) \times 10^3$	2
Total coliforms (#/100 ml)	9.0×10^5	1.3×10^3	3

Note: Loading rate average: 5 gal/day/sq ft (0.2 m/day). Numbers listed are mean values.

Concentrations of BOD_5 and TSS were significantly reduced by the sand filters. Almost complete nitrification of the septic tank effluent was also achieved by the sand filters. Orthophosphate concentrations were reduced approximately 20%. One to two log reductions of total and fecal coliforms were obtained through the sand filters; however, effluent concentrations will not meet current surface discharge recommendations of 1000/100 ml and 200/100 ml, respectively. Chlorination of the sand filter effluents reduced the coliform levels below the recommendations.

Aerobic Unit-Intermittent Sand Filter

Sand Clogging and Maintenance

A record of the infiltration rate of sands loaded with aerobic unit effluent was also maintained for the field and laboratory studies. Numerous curves were generated and a detailed discussion on the subject will be presented in a forthcoming publication.

Figure 2 shows the decline in infiltration rate with time when an aerobic unit effluent is applied to a sand. The data in this figure were generated

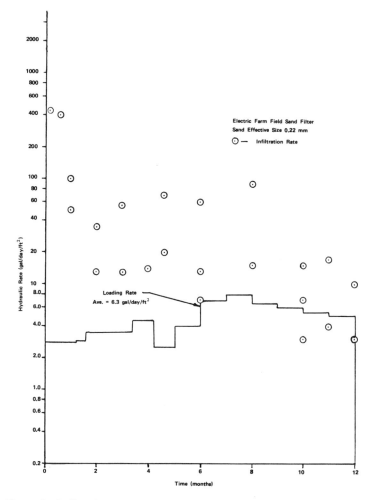

Figure 2. Infiltration rate decline of sand loaded with aerobic unit effluent.

at the Electric Farm field sand filter. The sand effective size was 0.22 mm, and the initial saturated hydraulic conductivity was 418 gal/day/sq ft (17 m/day). The average hydraulic loading rate was 6.3 gal/day/sq ft (0.25 m/day). The infiltration rate reduction in this figure represents the general pattern found in both the laboratory and field studies.

As shown in Figure 2, an immediate logarithmic decline in infiltration rate occurred during the first month of application. This was due to an accumulation of suspended solids, which were strained from the wastewater. This mat of accumulated solids formed on top of the sand surface and did not penetrate into the sand. The formation of the solids mat continued throughout the filter run, indicating a nondegradable nature of the suspended solids in the effluent of the aerobic treatment unit. Flow rates through the crust ranged from 50 to 100 gal/day/sq ft (2 to 4 m/day) at this time.

During the remaining filter run, infiltration rates ranged from 3 to 100 gal/day/sq ft (0.12 to 4 m/day). The range of infiltration rates was dependent upon the amount of time the crusted material on top of the sand remained unponded with wastewater. When continuous ponding occurred, infiltration rates decreased to as low as 3 gal/day/sq ft (0.12 m/day); however, when the crust was allowed to dry and crack open, infiltration rates were as high as 100 gal/day/sq ft (4 m/day). Eventually, continuous ponding predominated causing infiltration rates to lower below the 5 to 6 gal/day/sq ft (0.20 to 0.24 m/day) loading rate. When wastewater ponding above the sand reached 12 inches (30.5 cm), the filter run was ended. At this time, an accumulation of suspended solids on top of the sand had become ¾ to 1½ inches (1.9 to 3.8 cm), in depth. It is important to note that, even at failure, the ultimate infiltration rate was $\geq$ 3.0 gal/day/sq ft (0.12 m/day).

Sand columns studied in the laboratory also showed an infiltration rate reduction, as shown in Figure 2. The total run length, however, was shorter due to a higher suspended solids concentration and higher hydraulic loading rates. Larger sand sizes offered little increase in total run lengths; however, they appeared to have a higher ultimate infiltration rate at the end of the filter run. Ultimate infiltration rates in the laboratory ranged from 3.2 to 9.4 gal/day/sq ft (0.13 to 0.38 m/day).

Various maintenance techniques were studied to regenerate the clogged sand filters and columns. Results showed that removal of the solids mat from the top of the sand surface restored the infiltrative capacity to 50 to 100 gal/day/sq ft (2 to 4 m/day). This capacity was approximately equal to the sand capacity after the initial logarithmic decline in infiltration rate at startup. This allowed succeeding filter runs to be made without the need for resting the sand bed, thus eliminating the need for alternate filters. Apparently little active biological decomposition occurs below the sand surface, as a result of the low soluble organic material in the aerobic unit effluent. This statement is true only when the sand surface remains aerobic and at a high oxidation reduction potential. It also assumes that the aerobic treatment unit is properly operated and maintained.

The length of filter runs appeared to be dependent upon the mass of suspended solids applied to the sand surface. The mass is determined by the hydraulic loading rate and the concentration of suspended solids in the aerobic unit effluent. At loading rates ranging from 4 to 6 gal/day/sq ft (0.16 to 0.24 m/day) and suspended solids concentration $\simeq$ 50 mg/l, filter run lengths of 1 year have been experienced. At the same loading rate and suspended solids concentrations $\simeq$ 100 mg/l, filter run lengths of 6 months have been experienced.

Based on these findings, a hydraulic loading rate of 5 gal/day/sq ft (0.2 m/day) is recommended for determination of the required surface area of the sand filter. Maintenance to the sand filter is recommended at 6-month intervals, and involves the removal of the solids mat from the top of the sand surface along with 1 inch of sand. After replacing 1 inch of top sand, wastewater may be reapplied to the filter. No alternate filter appears to be necessary, if the aerobic treatment unit is properly operated and maintained. Periodic biological and hydraulic upsets can be assimilated by the sand filters; however, extended time periods of these upsets have led to shorter filter runs.

Effluent Quality

A detailed analysis of the effluent quality from sand filters loaded with aerobic unit effluent at an average rate of 3.8 gal/day/sq ft (0.15 m/day) has been published previously.[4] A few selected parameters are presented in Table VII. Further data from sand filters loaded at an average rate of 6.3 gal/day/sq ft (0.25 m/day) have shown similar results.

An important finding to note is that there is little difference in the effluent quality of the aerobic unit-sand filter effluent and the septic tank-sand filter effluent listed in Table VI. This is especially true after chlorination of the sand filter effluents, which reduced the coliform levels below recommended standards for primary contact recreational waters.

Disinfection Alternatives

Dry-Feed Chlorinators

The evaluation of commercially available dry-feed tablet-type chlorinators was performed at two home sites located on university experimental farms.

Table VII. Aerobic Unit-Sand Filter Effluent Quality Data

	Aerobic Unit Effluent	Sand Filter Effluent	Chlorinated Effluent
BOD_5 (mg/l)	26	2-4	4
TSS (mg/l)	48	9-11	7
Ammonia-N (mg/l)	0.4	0.3	0.4
Nitrate-N (mg/l)	33.8	36.8	37.6
Orthophosphate (mg/l)	28.1	22.6	23.4
Fecal coliforms (#/100 ml)	1.9×10^4	1.3×10^3	8
Total coliforms (#/100 ml)	1.5×10^5	1.3×10^4	35

Note: Loading rate average: 3.8 gal/day/sq ft (0.15 m/day). Numbers listed are mean values.

Experiments involved the chlorination of sand filter effluent over an 18-month time period. Collected sand filtered effluent flowed by gravity through a dry-feed chlorinator, and finally into a chlorine contact chamber. The filtered effluent dissolved the tablets in the chlorinator as it trickled past the tablets, while the contact chamber provided detention times ranging from 3 to 21 hours.[6]

Total residual chlorine concentrations are equally important in the evaluation of chlorinator performance. A range of total residual chlorine concentrations of 0.1 to 1.0 mg/l were measured after 4.5 to 18 hours of contact time. Because total residual chlorine concentrations > 0.05 mg/l can be toxic to aquatic plant and animal life,[7] acceptable discharge of this effluent is dependent upon the dilution effects of the receiving water.

A hypochlorite tablet uptake rate of 0.6 to 0.9 tablets/1000 gal (3.8 m^3) was measured for the aerobic unit-sand filter effluent at a flow rate of 100 to 150 gal/day (0.38 to 0.56 m^3/day). For the septic tank-sand filter effluent at a flow rate of 280 gal/day (1.06 m^3/day), a rate of 1.0 tablet/1000 gal (3.8 m^3) was found. Using these uptake rates, an approximate cost of chemical for chlorination was determined. For flow rates of 100 to 280 gal/day (0.38 to 1.06 m^3/day) the cost would range from $15 to $54 per year.

A major problem associated with the use of dry-feed chlorinators was the lack of control of the hypochlorite dose to the wastewater. This was a function of the periodic improper dissolving of the hypochlorite tablets. Various operation and maintenance techniques have been suggested to correct this problem.[6]

Ultraviolet Light Irradiation

Initial studies on the use of ultraviolet (UV) irradiation as a method of disinfection for sand filter effluent have also been performed. A commercially available UV water purifier was installed and is being tested at a home site located on a university experimental farm. Experiments have examined the UV irradiation of aerobic unit-sand filter effluent and of septic tank-sand filter effluent. The capacity of the UV water purifier was specified at 4 to 16 gal/min; however, initial tests were performed at a rate of 4 gal/min based on previous laboratory studies. The unit was also equipped with automatic wipers, which periodically clean the quartz jacket surrounding the UV lamp. In an attempt to arrive at a simple and practical unit, the cleaning system was not used in the study.

The UV lamp was operated continuously; however, sand filter effluent was pumped through the unit on an intermittent batch basis dependent upon the flow rates from the sand filters. A summary of 4 months of operating data are presented in Table VIII. During the initial 2 months, UV irradiation of aerobic unit-sand filter effluent was performed, while in the latter 2 months UV irradiation of septic tank-sand filter effluent was performed. Results from these tests show that no detectable numbers of fecal or total coliforms per 100 ml of UV effluent were present.

Table VIII. Coliform Analysis of Sand Filter and UV Water Purifier Unit Effluent

	Aerobic Unit-Sand Filter Effluent Loading Rate = 2.0 gal/day/sq ft (0.08 m/day)	UV Effluent	Septic Tank-Sand Filter Effluent Loading Rate = 3.2 gal/day/sq ft (0.13 m/day)	UV Effluent
Fecal coliform (#/100 ml) mean	11-13	0	$(2.6\text{-}4.4) \times 10^3$	0
Total coliform (#/100 ml)	64-75	0	$(3.6\text{-}5.1) \times 10^3$	0

It should also be noted that cleaning of the quartz jacket surrounding the UV lamp has not been required in 4 months of operation. Power requirements to operate the lamp continuously have been approximately 1.5 kwhr/day. At 4¢/kwhr, this represents an operational cost of approximately $22/yr.

A large drawback in the use of UV irradiation as a method of disinfection for small onsite wastewater systems is the large initial capital investment. From the initial results of this study, it appears that a much smaller and simpler UV unit could be built to adequately disinfect sand filter effluent. By doing this, initial capital costs could be reduced considerably, making the process more economically attractive.

Cost Analysis

The cost-effectiveness of any onsite wastewater treatment and disposal system is an important consideration. For the surface discharge disposal system proposed here, the costs are largely dependent upon the amount of wastewater to be treated, the availability of quality filter sand, and the amount of maintenance required by the system. A cost analysis involving the application of septic tank effluent and aerobic unit effluent onto sand filters has been performed. A summary of this analysis is presented in Table IX. Assumptions in the analysis include a three-bedroom

Table IX. Initial Capital Costs and Annual Operation and Maintenance Costs

Unit	Cost, in Dollars
Septic tank (1000 gal)	
Equipment and installation cost	350-450
Maintenance cost	10/yr
Aerobic treatment unit	
Equipment and installation cost	1300-2000
Maintenance cost	35/yr
Operation cost, 4 kwhr/day at 4¢/kwhr	60/yr
Wet well pumping chamber	
Equipment and installation cost	250-350
Operation cost,[a] ¾ kwhr/day at 4¢/kwhr	4/yr
Sand filter	
Equipment and installation cost	10-15
Maintenance cost	1
Chlorination and settling chamber	
Equipment and installation cost	700-1000
Operation cost[a] (chemical)	40/yr
Ultraviolet irradiation unit	
Equipment and installation cost	1100-1500
Operation cost,[a] 1½ kwhr/day at 4¢/kwhr	20/yr
Maintenance cost, cleaning and lamp replacement	undetermined

[a] Does not include pump replacement.

home, a family size of five, wastewater production of 50 gal/capita/day (0.19 m^3/capita/day) and the availability of a sand with effective size $\simeq$ 0.4 mm and uniformity coefficient of $\simeq$ 3.5. It is noted that sampling costs are not included in the cost analysis. Because discharge is to surface waters, state regulatory agencies may require some type of monitoring program.

The cost ranges presented in Table IX suggest that the two alternatives examined in this paper have similar although high costs when compared with septic tank-soil absorption fields, or with sewered waste treatment systems for small communities. These costs could likely be reduced if water conservation were practiced and, quite possibly, if waste segregation were employed. It must be recognized, however, that isolated systems can be evaluated only on a case-by-case basis, and conclusions on cost-effectiveness cannot be drawn by examining national averages.

APPLICATION OF SURFACE DISCHARGE TREATMENT SYSTEMS

Individual Home Sites and Community Systems

Although it has been shown technically that a high-quality effluent can be produced at individual home sites, numerous regulation, maintenance and institutional problems must be solved before widespread use of such systems can occur.[8] Unlike conventional subsurface disposal systems, surface discharge systems must be concerned with a point of final disposal. Whether the disposal point is a stream, lake, ditch, underground drain tile or an open field, potential environmental and health problems must be considered. These problems are in turn dependent upon the number of systems within a prescribed area. Another problem is to insure the performance of maintenance required for proper operation of the treatment system.

There is, however, a rational approach to the solution of these problems. The first step would involve the installation of a few experimental systems. These systems would be maintained by an installer or other trained personnel, bonded and under contract with the homeowner. Monitoring and regulation of surface discharge systems should be performed by some regional or local governing agency. Such agencies may include state regional offices, county offices or sanitary districts. These people would insure that the system is operating properly, and that proper maintenance is being performed. Undoubtedly, unforeseen management problems will occur during the experimental period. Solutions to these problems can be determined only if experimental systems are installed and tested.

If step one proves successful, regulatory officials should carefully outline a policy on the future use of surface discharge systems. Because of the relative complexity of such systems, a suggested policy outline includes a basic set of regulations, followed by a case-by-case approval. The basic set of regulations should specify the following conditions and requirements:

1. The type of home where a surface discharge system may be used; e.g., an existing home, a seasonal home, new homes installed with low-flow fixtures, etc. (Sand filter systems are ideally suited for seasonal homes or homes installed with low-flow fixtures. The periodic use of seasonal homes would reduce the amount of periodic maintenance to the sand filters. Waste segregation and low-flow devices would reduce the probability of hydraulic upsets and lower the nutrient discharge concentrations of nitrogen and phosphorus.);

2. The number of homes having surface discharge systems allowable within a defined region; e.g., a housing complex surrounding a lake or a small town on a trout stream may have to be concerned with nitrogen, phosphorus and bacteriological concentrations; hence, only a limited number of systems may be allowable;

3. The procedure for properly designing the surface discharge system;

4. A contracted agreement between the homeowner and a separate entity to insure the system is properly operated and maintained;

5. The discharge requirements and the amount of monitoring required by the governing agency; and

6. The legal authority of the governing agency when violation of the regulations occurs.

A more detailed and comprehensive discussion concerning regulatory methods for onsite sewerage systems can be found in the literature.[8]

In conclusion, this paper has examined only one concept of onsite wastewater treatment to achieve surface water quality standards. The impact of in-house waste segregation, water conservation, product and appliance selection, and recycling will all be significant. In addition, the need for other treatment processes within the flow sheet to provide for nutrient removal or other effluent polishing will greatly alter the cost-effectiveness of the system. Many of these alternatives are currently being evaluated by SSWMP, and future reports will be forthcoming.

REFERENCES

1. Siegrist, R. and N. Hutzler. "The Manipulation of Household Wastewater," Small-Scale Waste Management Project, University of Wisconsin, Madison, Wisconsin (1975).
2. Stothoff, J. R. "The Effect of Applied Wastewater, Loading Rate and Sand Size on the Performance of Intermittent Sand Filters,"

Independent Study Report, Department of Civil and Environmental Engineering, University of Wisconsin, Madison, Wisconsin (1976).
3. Hutzler, N. J. "Aerobic Treatment of Household Wastewater," Small-Scale Waste Management Project, University of Wisconsin, Madison, Wisconsin (1976).
4. Sauer, D. K. "Intermittent Sand Filtration of Septic Tank and Aerobic Unit Effluents Under Field Conditions," M.S. Thesis, Department of Civil and Environmental Engineering, University of Wisconsin, Madison, Wisconsin (1975).
5. Sauer, D. K., W. C. Boyle and R. J. Otis. "Intermittent Sand Filtration of Household Wastewater," *J. Environ. Eng. Div. ASCE* 102:EE4, Proc. Paper 12295 (August 1976).
6. Sauer, D. K. "Dry Feed Chlorination of Wastewater On-Site," Small-Scale Waste Management Project, University of Wisconsin, Madison, Wisconsin (1976).
7. Zillich, J. A. "Toxicity of Combined Chlorine Residuals to Freshwater Fish," *J. Water Pollution Control Fed.* 44 (February 1972).
8. Stewart, D. E. "Regulatory Methods to Assure the Maintenance of On-Site Sewerage Systems," Paper 76-2031, 69th Annual Meeting of the American Society of Agricultural Engineers, Lincoln, Nebraska (June 28, 1976).

14

THE SEWAGE OSMOSIS CONCEPT FOR ONSITE DISPOSAL SYSTEMS—CLAY SOILS

Frank P. Coolbroth

> Frank Coolbroth-Sitton Septic Tank Co.
> Plymouth, Minnesota 55442

Earl Peterson from Minneapolis introduced me to the theory that water moves through a partition of wet clay when an electric potential is applied across it. This encouraged me to look into the possibilities of using the theory on drainfields. Mr. Peterson and I pursued this concept, and he installed the first two Sewage Osmosis Systems. Both systems were used in drainfields that had failed.

A great deal of research has been done in the past few years on septic tanks and drainfield systems. This has given us a better understanding of how these systems work, what causes them to fail, new construction methods, aerobic versus anaerobic bacteria, the importance of soil classification, etc. All this information is of great importance if we are to have workable systems that will protect our environment. Emphasis at one time was on the septic tank, so larger tanks were recommended. How good and efficient can you make a septic tank? Perhaps we should consider something to increase the lateral and upward movement of water through the soil. We could then have a smaller drainfield area. Substandard systems that have failed could, perhaps, be made to work again.

Presented here are some of the criteria that prompted us to use the Sewage Osmosis Concept; and which perhaps might make it easier for you to recognize some of the advantages of using this concept for onsite sewage disposal systems.

- Approximately 40% of soil clogging is caused by ferrous sulfide. Ferrous sulfide readily oxidizes to the soluble sulfate form when aerobic conditions are restored. The important factor in minimizing biological clogging of a soil seems to be the maintenance of aerobic conditions in the soil itself.[1]

- The bottom area of the percolation trench is essentially useless as an infiltrative surface; thus, the system's geometry permits the sidewalls to play their important role.

Soil is perhaps the best way to treat sewage, and some of the recent research data seem to indicate this to be true. The lateral movement of water through clay is a contributing factor for this treatment. This brings us to the Sewage Osmosis Concept.

- Electroosmosis was first observed by F. Reuss in 1809.[2] This investigator pushed two vertical glass cylinders into a mass of wet clay, filled the cylinders with water, and inserted metal electrodes. When an electrical potential was applied, the level of water rose at the negative, but fell at the positive pole to which suspended clay particles were attracted. These results were in accord with the view that the clay particles were negative with respect to the water. Electroosmosis has been applied to the removal of water from peat and moist clay.

- Electroosmotic dewatering is a method of drying out an excavated area. It is used to increase the strength of the ground, and permit tolerable excavation slopes. This is achieved by forcing the flow of groundwater to drainage wells by the application of a direct current between electrodes inserted in the ground. Electroosmotic dewatering is confined to fine-grained soils, such as silts and clays.

The groundwater between the fine grains is made up of three layers, an uncharged inner core, a negatively charged layer fixed to the soil grains, and a positively charged intermediate layer between the fixed layer and inner core. Normally the positive and negative water layers adhere to each other, but when an electric potential is applied between nearby electrodes, the positive layer will flow toward the negative electrode, dragging the uncharged water core with it.

- Accompanying the improved removal of sodium is a percolation brought by the release of hydrogen and other gases upward through the soil to improve the soil's permeability and friability.[3] For some time, scientists have employed electroosmotic phenomena. Such efforts embody the application of direct currents between anodes and cathodes established in fine-grain soils to increase underground water seepage. This electro-drainage effect found successful application by the Germans during and prior to World War II in reclaiming certain poorly drained areas for construction of railways and submarine pens.

In the case of fine-grain soils, which are normally incapable of being drained or, if drainable, only with great difficulty by usual methods, the phenomenon of electroosmosis is brought about in the capillaries of the soil. If the electrical current is flowing in the proper direction, the water particles are transported effectively through the pores of the soil in the direction of the lines of the electrical field to the cathode. As a result of this process, the increased movement of the underground water is far greater than that imposed by the natural hydraulic gradient.

By burying dissimilar materials on both sides of a drainfield, an electric potential will be developed in the soil water complex. This will occur without the use of any outside current, wiring, or power source, and is the basis for the sewage osmosis method.

One of our experiments consisted of excavating a trench 24 feet in length, 36 inches deep and 20 inches wide. A double layer of 6-mil polyethylene was used to line the entire trench. Plywood was inserted 24 inches from each end of the trench to leave room for the cathode and anode. Damp clay was tamped into the remaining 20 feet of trench. This clay previously had a percolation of 1½ inches in 24 hours. The plywood was then removed, and 4 inches of clay was added at the anode. This was covered with another layer of polyethylene, assuring us of no leaks when the mineral rock was added. For the test, water was added to the anode. The anode was 20 inches wide, 32 inches deep, and 24 inches long. Water totaling 4256.3 gallons was added to the anode for a period of 163 days, for an average of 26.1 gallons of water per day. The depth of water in the cathode varied depending on the amount of rainfall. The fact that the tight clay soil would accept the volume of water that was applied, and the fact that water appeared in the cathode, indicated that electroosmosis was increasing the lateral movement of water through clay. Evidence shows that large amounts of water can be raised through the soil by capillarity.

Sewage osmosis utilizes the natural electrical energy created by the saturated clay particles within the soil. The anode attracts negatively charged ions, such as oxygen, phosphates and nitrates. The positively charged hydrogen ion is attracted to the cathode, allowing the removal of the outer water layer from the clay particles.

As this outer water layer is removed, it is replaced by free water from the capillary waters, producing water movement toward the cathode. Hydrogen is attracted to the cathode and oxygen is attracted to the anode, helping to maintain aerobic conditions in the soil.

Samples from two Sewage Osmosis Systems were collected by Serco Laboratories, Roseville, Minnesota, and the test results indicated: a significant decrease in fecal coliform bacteria, nitrogen compounds had reached a stable end product of decomposition, and a significant increase in dissolved oxygen.

Iowa Sewage Osmosis, Inc. installed a Sewage Osmosis System for Iowa State University, Ames, in 1975. An identical system with no cathodes or anodes has also been installed for comparison between a standard system and a Sewage Osmosis System.

In actual practice, each installation of a Sewage Osmosis System must be designed to fit the individual job site—proper grading of the site, soil classification, amount of sewage, etc. After obtaining the required information, a system can be designed by using a cathode or cathodes comprised of two high-carbon materials. The drainfield trench is divided into segments of various lengths, depending on the amount of sewage going into the system, anodes installed in each segment of trench, on the opposite side of the trench from the cathode. Anodes are constructed with rock high in minerals and oxides. This creates approximately 0.7 to 1 volt of direct current. Each segment of trench must be level, using a minimum of 24 inches of washed drainfield rock under the tile line. Thus, we have water that will start to flow laterally toward the cathode. We are also utilizing more sidewall area for this water movement. Inspection and sampling pipes are installed at the end of each segment of trench. Homeowners can measure the depth of water in each segment and know how their system is performing. A maximum of 8 inches of earth-cover over the tile lines is a very important factor for the release of hydrogen and other gases upward through the soil; water is also raised up through the soil by capillarity.

The area required for most single-family dwellings is only 50 by 60 feet. The cost to install a Sewage Osmosis System will vary from area to area, depending on the availability of materials. In the surrounding area of Minneapolis and St. Paul, Minnesota, the average cost is $2500.

During the past few years, I have been told on several occasions that it is impossible; yet, I have installed close to 50 Sewage Osmosis Systems, with a large number installed where other systems had failed. It is very gratifying to receive a letter from a happy, satisfied homeowner.

Burying dissimilar materials on opposite sides of trenches to create an electrical potential between cathode and anode has many advantages:

1. Systems can be installed in soils with zero percolation.
2. Systems that have failed can be made workable.
3. In most cases, only a small area is required.
4. The homeowner has an opportunity to check the system for performance.
5. The only maintenance required is pumping the septic tank.
6. Sewage Osmosis utilizes the natural energy created by the saturated clay particles within the soil.

All states are having problems finding suitable soils for onsite disposal systems. As the need for homes increases to meet our growing population, soil problems for onsite systems will also increase. A gallon of water taken from our underground water supply and not replenished could lead to serious problems in the future. Workable onsite disposal systems are an excellent way to replenish that supply.

The Sewage Osmosis Concept for onsite disposal systems certainly deserves serious consideration. A schematic illustration of the system, and data from a typical installation are appended.

REFERENCES

1. McGauhey, P. H. and J. H. Winneberger. "Causes and Prevention of Failure of Septic Tank Percolation Systems," Technical Studies Report No. 533, pp. 22,27.
2. Reuss, F. F. "Sur un Nauvel Effet de l'electricite' galvanique," *Mem. Soc. Imp. Naturalistes*, McGraw-Hill Encyclopedia of Science and Technology, 2:327-337 (1809), pp. 525,526.
3. Collopy, J. P. "Improvement and Reclamation of Soils," Patent Application 2831804, April 22, 1958, Vol. II,III.

136 NSF THIRD NATIONAL CONFERENCE

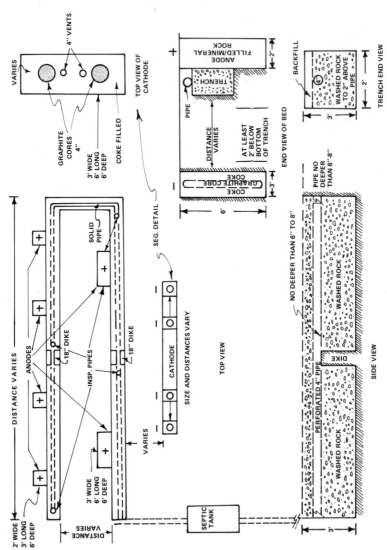

Twin City Water Clinic

14613 KARYL DRIVE • HOPKINS, MINNESOTA 55345 • 935-3556

J. R. Van Arsdale, Registered Professional Engineer

Coolbroth - Sitton Septic Tanks   May 6, 1975
4810 W. Medicine Lake Dr.
Minneapolis, Minn. 55442

REPORT OF WATER ANALYSIS

Our laboratory reports these analytical results, determined on a sample taken by us on May 1, 1975

        Howard Sedor
        8272 Peony Lane
        Maple Grove

	Cathode	#3 Inspection
Bacteria (coliform group)	0/100 ml	30,000/100 ml
Nitrate nitrogen	0.0 ppm	0.0 ppm
Surfactants (detergents)	0.10 inch	0.30 inch

The water from the cathode hole was relatively free of contamination from the sewage system. We did find a low level of surfactants. This water actually meets drinking water standards.

The #3 inspection hole shows contamination from the sewage system with coliform bacteria present and 0.30 ppm of surfactants (detergents) present.

                TWIN CITY WATER CLINIC

                J. R. Van Arsdale

Analytical laboratory	Consulting engineer
Water analysis reagents	Boiler water chemicals

.17.1 parts/million equals 1.0 grain/gallon

15

SOIL EVALUATION OF SITES FOR ABSORPTION SYSTEMS

Dale E. Parker

Soil Scientist
Wisconsin Department of Health and Social Service
Division of Health
Madison, Wisconsin 53701

INTRODUCTION

Successful treatment and disposal of liquid waste by a septic tank soil absorption system depends heavily on site conditions. Such dependence restricts the locations where these systems can be used. Because the septic tank can operate in a satisfactory manner almost anywhere, the real problem is the suitability of the site used for soil absorption. The factors considered in determining site suitability are: soil permeability, depth to seasonally saturated conditions, depth to bedrock, land slope and surface flooding hazard. Basically, a nearly level to gently sloping site free of flooding, with at least 5 feet of permeable soil, is needed for an absorption system.

A preliminary selection of sites can be made by determining general soil landscape characteristics, or by using detailed soil survey information. Onsite investigations, however, are necessary to determine soil characteristics by studying soil pits, to evaluate permeability by conducting percolation tests, and to observe other site factors.

PRELIMINARY EVALUATION

Landforms, soil landscapes and detailed soil maps provide general information useful in the selection of sites for absorption systems. Soil maps provide the greatest amount of information and should be used when available. When they are not available considerable information can be obtained from other resource maps showing broad landscape features, such as bedrock and glacial geology maps, or by airphoto interpretation of landforms and soil landscape units.

Landforms have particular characteristics, such as shape, mode of formation or deposition, and type of materials which indicate site conditions. Geologists, geomorphologists, soil classifiers and others can identify and separate different units of the landscape. The first-order separation might be into landforms, such as a glacial ground moraine, outwash plain or valley bottomland. This information can be used to make general estimates of the ability of different segments of the landscape to absorb liquids. For example, as shown in Figure 1, glacial ground moraines have long, gentle slopes; they consist of the finer, loamy and clayey textures; and they are usually more dense, resulting in slowly permeable conditions. Glacial

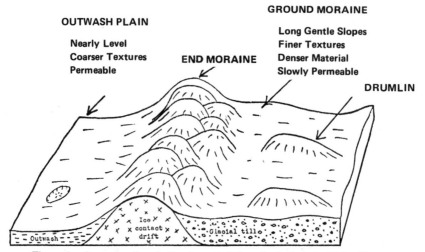

Figure 1. General soil and site characteristics of selected glacial landforms.

outwash plains, on the other hand, are nearly level, have coarser textures, and are likely to be rapidly permeable. Similarly, it is easy to see that representative residual soils weathered from limestone, shale or granitic bedrock will have absorptive qualities quite different from those weathered from a medium-grained sandstone. These few examples show the importance of landscape units in evaluating sites for absorption systems.

Detailed soil maps show the location, size and shape of different kinds of soil on the landscape. The different soils are usually delineated on airphoto base maps, and are called soil map units. The soil survey report describes the character of the map units and provides predictions or interpretations for their use and management. Soil map unit descriptions give the relative position in the landscape, the average land slope, the natural drainage class, and the characteristics of the soil profile. The different horizons of the soil profile are described in detail with regard to texture, structure and color, including any mottling. Soil profile descriptions also indicate other features significant to the operation of absorption systems, such as the layering of materials, and the depth and type of bedrock. Engineering test data and interpretations that rate the map units for liquid waste disposal are also provided in the survey report. Items of importance for site evaluation which can be obtained from the soil survey include: soil texture, structure and indicated permeability; natural drainage class and depth to mottling, which indicate the depth to seasonal saturation; depth to and type of bedrock; and, landscape position along with the land slope. A useful tool for site evaluation is a correlation of natural drainage classes with expected levels of saturation, as shown in Figure 2.

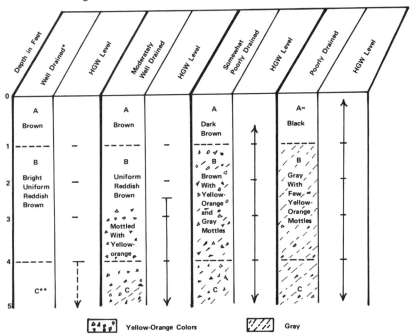

*Includes excessively drained shallow sandy soils lacking B horizon.
**May have few faint mottles in C horizon and HGW levels at less than 6 feet in some cases.

Figure 2. Schematic diagram of natural soil drainage groups for evaluating high groundwater (HGW) levels of soil absorption sites. On basis of soil pedology, high groundwater levels are determined to be at the uppermost level of prominent yellow-orange or grey mottling.

Land slope, as shown by the soil map or as observed in the field, is an important factor in estimating the absorptive capacity of an area. Convex slopes and steeper slopes shed water, remain dryer, and consequently are better able to absorb liquids. Concave slopes receive runoff and may occur as depressions which collect and hold water. These areas are frequently moist or even wet for long periods of time, and are less able to absorb more liquid. Areas with concave slopes may also be seasonally saturated at shallow depths.

Preliminary information obtained by landscape analysis or from soil maps permits the soil tester to anticipate soil conditions, make an early selection of the most desirable locations, and plan time and equipment needs for onsite investigations. Regulatory officials can also use the information for these purposes as well as for evaluating onsite test data reports to help decide the need for a field investigation prior to issuance of a sanitary permit.

FIELD EVALUATION

The field evaluation for an absorption system includes investigating landscape conditions, observing soil profile characteristics, and conducting percolation (or other) tests to determine soil permeability. Soil percolation test results are variable,[1-3] but they provide a means of ranking the soil's ability to accept liquids for use in system sizing. They cannot be used alone to adequately evaluate sites for soil absorption systems. Information on soil profile characteristics is necessary.

During the first observation of a site, specific areas which are expected to be suitable or unsuitable can frequently be selected. A view of the topography and vegetation will usually indicate the location of unsuitable conditions, such as wet soils, bedrock outcroppings, steep slopes, and areas subject to flooding by surface water. Areas which are obviously wet, are subject to frequent ponding or flooding, have rock outcroppings, or have slopes exceeding 20%, can immediately be considered unacceptable sites for soil absorption systems.

Soil bore holes, or preferably soil pits, are constructed in potentially suitable areas to evaluate soil profile characteristics before conducting percolation tests. Important characteristics to be considered are soil texture, structure, color patterns, layering of different textures and bedrock conditions. Soil texture, structure and layering characteristics are used to estimate soil permeability. In general, soils with coarse textures, strong granular, blocky or prismatic structure, and soil profiles free of textural layering are the most permeable. Soils which are massive, have a platy or weak structure, and have fine textures tend to be less permeable.

Soil color patterns are used to identify seasonally saturated soil layers.[4] Most subsoil colors result from the oxidation-reduction status of iron compounds. Well-drained soils have uniform bright-colored brown, reddish-brown or yellowish-brown subsoils because the iron compounds are oxidized. Poorly drained soils which are saturated most of the time have dull, grey-colored subsoils because the iron has been reduced or removed from the soil. Subsoils subject to alternating or seasonal periods of saturation are mottled with grey, brown and yellow-orange colors. Sufficient permeable soil is needed to maintain at least 3 feet of separation between the absorption system and seasonally saturated layers.

Bedrock, if solid or if cracked and fractured with open crevices, should also be separated from the absorption system by at least 3 feet of soil to provide reasonable assurance of removing bacteria and viruses.

If soil pits show sufficient permeable soil, essentially 5 feet, free of bedrock and seasonally saturated layers, soil percolation tests should be conducted at the depth of the proposed system. Percolation tests must be carefully conducted following a standardized procedure which provides for adequate soaking and swelling of medium- and fine-textured soils. Testing should continue until a stabilized rate is achieved. Rates of less than 60 minutes for the water to fall 1 inch are generally considered adequate. The percolation rate obtained is used in sizing the soil absorption system.

In addition to the previous factors discussed for evaluation of soil absorption sites, adequate separation from lakes, streams, wells, water lines and buildings must be maintained.

Soil and site factors for absorption systems should be evaluated either by specially trained local code administrators or licensed private contractors. In Wisconsin, sites are evaluated by specially trained and licensed certified soil testers.[5] When the soil and site information indicate suitable conditions, the data are used in designing the soil absorption system.

NEW TECHNIQUES

Several new techniques for evaluation of soil absorption system sites have been suggested. Bouma[6] suggested using the crust test to determine hydraulic conductivity rates for benchmark soils along with greater use of soil classification and soil maps to extend the information from one site to the next. To follow his suggestions, key properties important to liquid waste disposal are measured and determined for a particular kind of soil. This is done by testing several sites to determine the range of key factors within a particular soil. The information obtained is used by transferring it to other sites with similar soils as determined by taxonomic classification.

If the new site is classified the same, similar key properties can be expected and repetitive, time-consuming, onsite tests to determine absorptive qualities of the soil are not needed. This method assumes that the taxonomic classification adequately reflects key soil properties. In most cases this is true, and soil variability is sufficiently low to permit direct evaluation of sites by this method. It also assumes availability of trained staff, or licensed contractors, capable of proper soil classification.

Soil maps can also be used to extrapolate important information. When the new site is shown by the soil map to be in a mapping unit named for the same soil series, similar key properties can be expected. This is less dependable than onsite observation and classification because of inclusions in soil map units and scale limitations. However, soil maps can be used where the map units are homogeneous, have few inclusions, and a low variability of key properties.

Lake County, Illinois is using a combination method of site evaluation which has been successful.[7] This method uses estimated percolation rates for the soil series shown on soil maps. Onsite evaluations are used to verify the soil shown on the map, or to make the proper classification. With qualified staff and use of a soils key, good results have been obtained without conducting time-consuming percolation tests.

These new techniques for evaluating sites have been used in varying degrees throughout the country. They are still being evaluated and improved by research.[8] While clear-cut success in all cases is not evident, most of the techniques do show promise for future use in evaluating sites for soil absorption systems.

SUMMARY

This paper includes a discussion of the factors which are important in the evaluation of sites for absorption systems. While many factors are important in making an evaluation, key steps include the use of detailed soil maps and soil pits for determining soil profile characteristics. It should be clearly evident that soil percolation tests alone do not provide adequate data for site evaluation.

REFERENCES

1. Bouma, J. "Evaluation of the Field Percolation Test and an Alternative Procedure to Test Soil Potential for Disposal of Septic Tank Effluent," *Soil Sci. Soc. Am. Proc.* 35:871-875 (1971).
2. Derr, B. D., R. P. Matelski and G. W. Peterson. "Soil Factors Influencing Percolation Test Performance," *Soil Sci. Soc. Am. Proc.* 33:942-946 (1969).

3. Hill, D. E. "Percolation Testing for Septic Tank Drainage," Connecticut Agricultural Experiment Station Bulletin No. 678, 25 pp. (1966).
4. Veneman, P. L. M., M. J. Vepraskas and J. Bouma. "The Physical Significance of Soil Mottling in a Wisconsin Toposequence," *Geoderma.* 15:103-118 (1976).
5. Hill, R. C. and D. E. Parker. "Soil Testing for Liquid Waste Disposal in Wisconsin," *Proceedings of Illinois Private Sewage Disposal Symposium,* Champaign, Illinois, pp. 27-33 (1975).
6. Bouma, J. "New Concepts in Soil Survey Interpretations for Onsite Disposal of Septic Tank Effluent," *Soil Sci. Soc. Am. Proc.* 38:941-946 (1974).
7. Mellen, W. L. "Identification of Soils as a Tool for the Design of Individual Sewage Disposal Systems," Lake County Health Department, Waukegan, Illinois. 67 pp. (1976).
8. Baker, F. G. and J. Bouma. "Variability of Hydraulic Conductivity in Two Subsurface Horizons of Two Silt Loam Soils," *Soil Sci. Soc. Am. J.* 40:219-222 (1976).

16

SEPTAGE DISPOSAL IN WASTEWATER TREATMENT PLANTS

Ivan A. Cooper and Joseph W. Rezek
Rezek, Henry, Meisenheimer and Gende, Inc.
Consulting Engineers
Libertyville, Illinois 60048

GENERAL

The first priority of the Environmental Protection Agency's program to abate water pollution has been focused on providing adequate wastewater treatment for sewered communities. However, results of the 1970 census inform us that 16.6 million housing units, or over 24.5% of the total housing units in the United States, relied on septic systems for wastewater disposal.

Users

The geographical distribution of the use of septic systems, as seen in Figure 1, shows states with over 35% usage located in New England, the Southeast, and the Pacific Northwest. Most North Central, Northeastern and Southeastern states have only a slightly lower usage of these onsite disposal facilities. The Southwest states' usage of septic tanks is between 10 and 20%. On a local level, many counties in New Jersey, New York, California and other states have over 50,000 housing units which use onsite waste disposal systems, while their statewide usage appears less significant. Areas with over 100,000 housing units using onsite waste disposal systems include suburban New York, Los Angeles and Miami.[1]

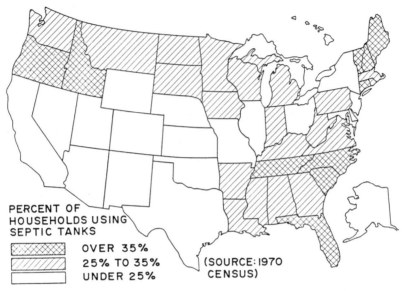

Figure 1. Geographical distribution of the use of septic systems.

The use of a septic system requires periodic maintenance which includes pumping out the accumulated scum and sludge, called septage. Kolega[2] has reported a septage buildup of from 65 to 70 gallons per capita per year in properly functioning septic systems.

Various recommendations exist for time periods between pumping out a septic tank—most, between two and five years. After a hauler pumps out the homeowner's septage, this highly offensive sludge must be disposed of in a safe, cost-effective and convenient manner.

Septage Characteristics

Septage is a highly variable anaerobic slurry with characteristics which include large quantities of grit, grease, highly offensive odor, the ability to foam, poor settling and dewatering, high solids and organic content, and quite often, an accumulation of heavy metals. Tables I, II and III are the result of previous research work compiled by the U.S. EPA's Cincinnati research group, as well as extreme values reported in the literature.

Graner[3] reports septage characteristics in Nassau and Suffolk counties as low as medium to strong wastewater, while Goodenow[4] in Maine found some samples with total solids and suspended solids over 130,000 mg/l and 93,000 mg/l, respectively. Tilsworth[5] in Alaska obtained some septage

Table I. Septage Characteristics (all values in mg/l except where noted)

Parameter	EPA Mean Concentration	Minimum Reported	Maximum Reported
Total solids	40,000	1132[3]	130,475[4]
Total V.S.	26,000	4500[7]	71,402[4]
Total S.S.	15,000	310[5]	93,378[4]
Volatile S.S.	18,100[7]	3660[7]	51,500[9]
BOD_5	5,000	440[3]	78,600[5]
COD	45,000	1500[5]	703,000[5]
TOC	15,000	1316[6]	96,000[8]
TKN	600	66[6]	1,900[7]
NH_3-(N)	150	6[6]	380[8]
NO_2	0.7[7]	< 0.1[8]	1.3[8]
NO_3	3.2[7]	< 0.1[8]	11[10]
Total P	150	20[7]	760[6]
PO_4	64[7]	10[7]	170[7]
Alkalinity	1,020[7]	522[5]	4,190[5]
Grease	9,561	604[6]	23,368[6]
pH (units)	6-9	1.5[3]	12.6[3]
LAS	150	110[6]	200[6]

Table II. Septage Metal Concentrations (all values in mg/l)

Parameter	EPA Mean Concentration	Minimum Reported	Maximum Reported
Al	50	2[6]	200[6]
As	0.1	0.03[6]	0.5[6]
Cd	0.5	<0.05[6]	10.8[6]
Cr	1.0	0.3[6]	3.0[8]
Cu	8.5	0.3[8]	34[6]
Fe	200	3[6]	750[6]
Hg	0.1	0.0002[6]	4.0[6]
Mn	5.0	0.5[6]	32[6]
Ni	1.0	0.2[6]	28[8]
Pb	2.0	1.5[6]	31[6]
Se	0.1	<0.02[6]	0.3[6]
Zn	50	33[8]	153[6]

Table III. Heavy Metal Content of Septage and Municipal Sludge[6] (mg/kg)

Metal	Lebanon, Ohio Septage	Lebanon, Ohio Salotto	Other U.S.	Denmark	Sweden
Cd	5.5	43	69	10	9.3
Cr	21.0	1050	840	110	170
Cu	28.1	1270	960	340	670
Hg	<0.14	6.5	28	7.8	5.8
Mn	106	475	400	350	400
Ni	<23.5	530	240	37	65
Zn	1280	2900	2600	2600	1900

samples with BOD_5 over 78,000 mg/l and CODs over 700,000 mg/l. The EPA's mean concentrations are good indicators of septage concentrations when compared with other researchers' data.

The geometric mean heavy metals content of residential septage from Lebanon, Ohio was compared with geometric means found in raw and digested sludge from 33 U.S. sewage treatment plants, and Danish and Swedish sludge metals content as shown in Table III. On a mg/kg dry weight basis, domestic septage contains one-half to two orders of magnitude less heavy metals than does municipal sludge.[6]

Bacteriology

Bacteriologically, septage contains predominantly gram-negative, nonlactose fermenters. Many of these microorganisms, such as *Pseudomonas*, are considered aerobic and have been found in septic tanks. Numerous obligate anaerobes are present, but only spore-forming types including *Clostridium perfringenes* have been recovered.

When the septic tank is pumped, mixing of the bottom sludge, intermediate wastewater and upper layer of scum occurs, yielding both aerobes and obligate anaerobes. The presence of aerobic types in a septic tank can be explained by the dissolved oxygen of the incoming sewage providing sufficient oxygen to allow limited aerobic growth, or displacement of effluent by the influent furnishes a relatively constant number of aerobic microorganisms.[11]

Calabro[11] estimated the relative stability of septage, septic tank sewage, and domestic wastewater using methylene blue as a redox indicator of biological activity. Septage samples changed color in 5 hours, septic tank sewage between 6 and 21 hours, and raw domestic sewage between 17 and 21 hours.[11]

Numerous alternatives are currently available for septage disposal. Although this paper is directed toward disposal practices in a wastewater treatment facility, alternatives in the broad categories of land disposal and separate treatment facilities will be outlined.

LAND DISPOSAL

Septage disposal on the land can include surface spreading and subsurface injection, spray irrigation, trench and fill, sanitary landfills and lagooning. Common requirements in all land disposal alternatives are analyses of soil characteristics, seasonal groundwater levels, neighboring land use, groundwater and surface water protection and monitoring, climatological conditions, and site protection, such as signs and fencing.

Landspreading requires a knowledge of land slopes, often limited to 8%, and runoff conditions; storage facilities for times when land application is inadvisable, such as during rainfall periods when the land is frozen or has a snow cover; crop management techniques; odor control procedures, which may include discing or lime spreading; and loading criteria. Loading criteria include hydraulic, organic and heavy metals limitations. The lower heavy metals content of septage compared with municipal sewage sludge will allow a 5 to 8 times greater application rate than municipal sludge if heavy metals are the limiting criteria.[6] The State of Maine published an excellent report, *Maine Guidelines for Septic Tank Sludge Disposal on the Land*, which addresses loading criteria.

Designers of spray irrigation sites should have knowledge of prevailing wind patterns for odor control, and use storage lagoons for times when spraying is not advisable, and a screening device on the lagoon's pump suction to prevent clogging at the distribution nozzles. Spray irrigation fields should be plowed under when an area's use is terminated.

Septage disposal in trenches is similar to disposal in lagoons, except that trenches are usually a smaller-scale alternative than the lagoon. Septage is placed sequentially in one of many 7-foot maximum depth trenches in small lifts, 6 to 8 inches, to minimize drying time. When the trench is filled with septage, 2 feet of soil is placed as a final cover, and new trenches are opened. The trench and fill technique is quite often used at sanitary landfills.

When a sanitary landfill accepts septage, leachate production and treatment must be investigated. For moisture absorption, New Jersey recommends a starting value of 10 gallons of septage to each cubic yard of solid wastes. A 6-inch earth-cover should be applied daily to each area that was dosed with septage.[12]

Different types of lagoons have varying acceptability as a septage disposal method. Aerobic lagoons are not recommended, because each

1000-gallon septage discharge would use between 1 and 3 acres of surface area, based on a design of 20 to 50 pounds BOD/acre/day.

Anaerobic Lagoons

Anaerobic storage or septage lagoons operate best when lime addition is used to maintain the pH between 6.8 and 7.2, and a minimum 20-day retention period is provided at maximum hydraulic loading.[12] Series or series-parallel lagoons with 2 years capacity each and 2-foot maximum depth are called for in New York State Guidelines.[13] Disposal lagoons are usually a maximum of 6 feet deep and allow no effluent or underdrain system. These disposal lagoons require small (6 to 12 inches) application rates and sequential loading of lagoons for optimum drying. After drying, solids may be bucketed out for disposal in a sanitary landfill with use of the lagoon for further applications—or 2 feet of soil placed over the solids as a final cover. Odors may be controlled by placing the lagoon inlet pipe below the liquid level, and clean-up water available for haulers to immediately wash any spills into the lagoon inlet line.

SEPARATE TREATMENT FACILITIES—SEPTAGE ONLY

Alternatives for treating septage at a separate treatment facility include aerated lagoons, anaerobic/aerobic processing, composting, the BIF Purifax process, and chemical treatment.

Aerated Lagoon

Aerated lagoons may be employed for treating septage if the aerators have the required oxygen transfer capacity and impart sufficient turbulence to prevent solids deposition. Howley[14] reported severe foaming problems, but he did obtain a VSS reduction of 23.8% and a COD reduction of 73.9% using various hydraulic retention times of 1 to 30 days. He found 1.8 pounds of oxygen were required to destroy 1 pound of VSS at loadings between 0.03 and 1.3 pounds VSS/ft^3/day, compared with an accepted 1.42 pounds of oxygen per pound VSS destroyed in raw domestic sewage,[14] and reported that 18,500 gpd/million gallons of aerated lagoon design capacity operating at 50% design sewage flow should not cause an overload condition.

Brookhaven, Long Island, using lagoon treatment of septage, experienced reductions of 62.5% in BOD, 51% in total solids, 49% in suspended solids from influent strengths averaging 5600, 3700 and 2700 mg/l, respectively. Without equalization facilities, this process was prone to biological upsets. Grit and scum chambers and three large settling lagoons now buffer flow to the 50,000-gpd septage system. The effluent from a final settling lagoon is chlorinated and discharged to sand recharge beds. Accumulated sludge is removed to a nearby landfill.

Anaerobic/Aerobic Process

The anaerobic/aerobic process uses an anaerobic lagoon or digester prior to an aerated lagoon. A pilot plant anaerobic/aerobic treatment process with sand beds for filtering final effluent reported 99% BOD, COD and SS removal, and 90% removal of total nitrogen and total phosphorus.[15] Anaerobic digesters are useful in reducing concentrations of volatile solids and BOD, and are covered later in this paper.

Composting

Composting is an alternate septage disposal technique offering a potential for good bactericidal action[16,17] and a 25% reduction in organic carbon. The Lebo and the Beltsville processes appear promising. The Lebo method uses a patented preaeration process prior to spraying septage on sawdust piles. A 1- to 2-inch application is covered with additional sawdust, and the mixture is formed with front-end loaders into piles to minimize heat loss. Natural draft aeration is possible because of the bulky nature of this mixture, eliminating the need for turning or forced aeration. Three months composting the 50-60% moisture content material at a pile temperature of 150°F produces a saleable landscaping material. The Beltsville system mixes sludge with wood chips in long windrows, and has piping facilities to alternately blow and pull air through the windrows to maintain aerobic bacterial action.[16,17] Some turning of the windrows is suggested. After several weeks, the compost can be screened, recycling the wood chips for further composting. Both systems should have ample room on site for movement of heavy equipment, as well as facilities for collection of leachate and surface water. Primary screening for removal of larger unwanted material is advised.

Purifax

The BIF Purifax process oxidizes screened, degritted and equalized septage with dosages of chlorine from 700 to 3000 mg/l under moderate pressure. Chlorine replaces oxygen in organic molecules, rendering this material unavailable to bacteria as a food source, thereby stabilizing and deodorizing the septage. The purifaxed septage changes color from black or deep brown to a straw color. The process initially releases CO_2 gas which separates liquids and solids quickly, with rising solids accumulating for removal.

Purifax treatment results in a highly acidic slurry, pH 1.7 to 3.8. If mechanical dewatering or lagoon separation of the liquids or solids is contemplated, chemical addition for pH control of the resultant liquid fraction must be included.

A Purifax treatment scheme at Lebanon, Ohio pressure-chlorinated septage and then dewatered the processed septage on sand drying beds. Analyses of underdrainage showed LAS, COD, phosphorus and iron removals of 99%, BOD_5 removal of 97%, zinc removal of 96%, and nitrogen removal of 83%.

Only 1 to 2½ days were required to dry a 6- to 12-inch application to 30% solids. The sand bed underdrainage has been measured at a pH 6.8 to 7.2. Residual organic chlorine was found only in the dry solids, in the range of 1.6 to 3.5% by weight.[6]

Other locations which use the Purifax process treating septage and/or sludge with lagoons for liquid-solid separation have had periodic solids separation and odor problems. Sand drying beds appear to be the most efficient method of liquid-solids separation. Adequate ventilation of covered sand drying beds is mandatory to eliminate operator health hazards from inhalation of any NCl_3 released subsequent to the Purifax process.

Chemical Treatment

Raw septage is chemically treated with lime and ferric chloride at an Islip, Long Island facility. After screening, degritting and equalization, about 190 pounds lime/ton dry solids and 50 gallons/ton dry solids of a standard strength ferric chloride solution is flash-mixed with the septage. The solids-liquid separation step occurs in a clariflocculator. An observed significant solids carry-over problem indicates the separation unit may have been undersized. The liquid fraction is chlorinated and discharged to groundwater recharge beds. The underflow solids from the clariflocculator are vacuum-filtered. Long-term relative stability of the lime-ferric chloride-septage mixture is unknown.

Tilsworth[5] found good liquid-solids separation only with huge additions of chemicals—for example either 10,000 mg/l lime, 10,000 mg/l ferric sulfate, 4,000 mg/l lime and ferric sulfate mixture, or a 3% concentration of a cationic polymer.

SEPTAGE–SEWAGE TREATMENT FACILITIES

General

Septage can be disposed of in a treatment facility by addition to either the liquid stream or the sludge stream. In either case, screening, degritting and equalization are recommended.

Septage frequently is considered a high-strength wastewater and is dumped into an upstream sewer or placed directly into various unit processes in a treatment plant (Figure 2). In several facilities, septage is

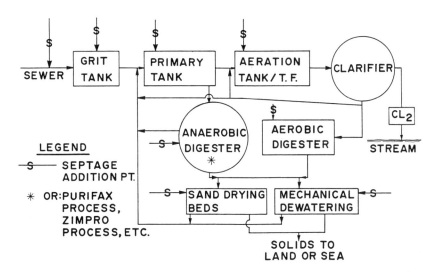

Figure 2. Septage addition points in wastewater treatment facilities.

considered a sludge because it is the product of an anaerobic settling/digestion tank, and is approximately the same total solids concentration as raw municipal sludge. The septage application points, if treated as a sludge, may include sludge stabilization, sand bed drying, or the mechanical dewatering process. The decision of where to apply the septage should be determined following a statistically significant sampling and analysis program of a locale's septage, including: solids loading, oxygen demand, toxic substances, foaming potential, and nutrient loading (N and P), where required.

These factors, combined with a plant's layout, design capacity, present loading, and appropriate design criteria provide the design professional with sufficient information to arrive at a reasonable septage treatment scheme within a wastewater treatment facility.

When septage is added to an upstream sewer or discharged at a treatment plant, a suitable hauler truck discharge facility should be provided. This facility should include a hard-surfaced sloping ramp to an inlet port to accept a quick disconnect coupling directly attached to the hauler's truck outlet. This will significantly reduce odor problems. Washdown water should also be provided for the hauler so spills may be cleaned up. A recording of the time, volume and name of the hauler is vital for both operation and billing purposes. Portland, Oregon's Columbia Avenue plant and Seattle's Metro Renton facility use a plastic charge plate or magnetically coded card and card reader.

Pretreatment

Treatment plants handling septage have experienced better operation when septage pretreatment is employed. Pretreatment generally includes bar screens of ¾- to 1-inch opening, grit removal, and preaeration or prechlorination if added to an aerobic process. Grit removal by cyclone clarifiers has been used successfully in Babylon, and is included in the new Bay Shore plant, both on Long Island, New York. Equalization/storage tanks of two days average septage flow and mixing capability should also be provided. To further attenuate odors, enclosed storage tanks and ozonation of tank vent lines may be considered. Pumping equipment should be used to apply a continual small dose of septage into the desired unit process. Operators report that slug doses or intermittent doses of septage are not as effective as a continuous feed rate.

Primary Treatment

Related to the sedimentation of septage, a report by Feiges[8] for the U.S. EPA indicated that neither natural settling, lime addition, nor polyelectrolyte addition resulted in consistent separation. Tilsworth[5] characterized raw septage as relatively nonsettleable as determined by settleable solids volume tests, from 0 to 90%, with 24.7% as the average volume.

Tawa[18] found septage settling characteristics could be divided into three groups, Types 1, 2 and 3. Type 1, from septic tanks pumped before necessary, settled well. Type 1 septage was found in 25% of his samples. Type 2 septage, from normally operating systems, showed intermediate settling characteristics and was found in 50% of his samples. Type 3 septage exhibited poor settling, was found in 25% of his samples, and was from tanks overdue for pumping. It was generally found that poor settling characteristics can be expected from septage.

Activated Sludge

Septage may be added to the activated sludge process if (1) additional aeration capacity is available, (2) the plant is hydraulically loaded below design capacity, (3) the septage metals contents can be diluted to a sufficiently low concentration, and (4) foaming potential is low or can be controlled. Very limited quantities of septage may be added without changing the sludge wasting rates.

At the Weaverville Waste Water Treatment Plant in Trinity County, California, 400-gpd slug dumps could be handled at a 0.5-mgd plant flowing at 40% capacity. The use of slug-dumping septage may depend on limiting the increase in MLSS to 10% per day to maintain a relatively stable sludge [19] as seen in Figure 3. Higher loadings and wasting rates than the resident aquatic biomass[20] is acclimated to may result in a poor settling sludge. Severe temporary changes in loading beyond the 10 to 15% MLSS increase may cause a total loss of the system's biomass.

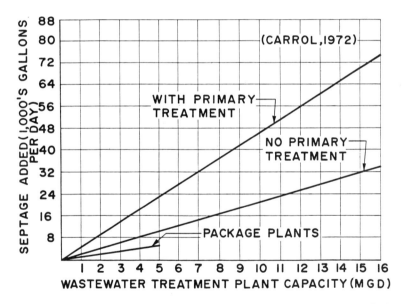

Figure 3. Septage addition to activated sludge wastewater treatment plants (no equalization facilities).

As seen in Figure 3, a lower level of septage input is required at a treatment plant lacking equalization facilities than at a plant having these facilities. A U.S. EPA study[6] fed septage at a controlled 2 to 13% flow rate to one of two activated sludge units. With a control unit food to microorganism ratio (F/M) of 0.4 and a septage-sewage F/M of 0.8, effluent BOD and SS characteristics were similar. Effluent COD of the unit receiving septage increased when septage was loaded at 10 to 13% of plant flow, and nitrification was retarded. When a lower F/M ratio of 0.5 to 0.6 was utilized in the septage unit, this unit had superior performance; Nocardia, a procaryotic filamentous antinomycete often associated with bulking, was controlled.

Figure 4 is indicative of continuous septage addition to a facility for a fully acclimated biomass. Initial septage feed to an unacclimated system

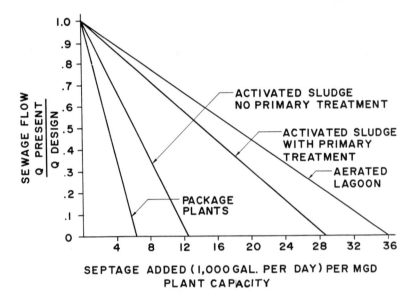

Figure 4. Septage addition to wastewater treatment plants with equalization facilities.

should be substantially less than shown on the graph, *i.e.*, on the order of 10% of the graph values. Further gradual increases should be made in the range of 5 to 10% of the current septage flow per day up to the maximum amount shown on Figure 4. Oxygen capacity must be checked continuously (Figure 5), and gradual changes made in sludge age.

Because of the higher oxygen demand for septage than for raw sewage on a unit BOD_5 basis, an additional oxygen supply for activated sludge plants accepting septage having primary treatment would be 40 pounds O_2/1000 gal septage added. For plants without primary treatment, an additional 80 pounds O_2/1000 gal septage added should be provided. Feng has shown that higher sludge ages (10 days vs. 4 days) result in

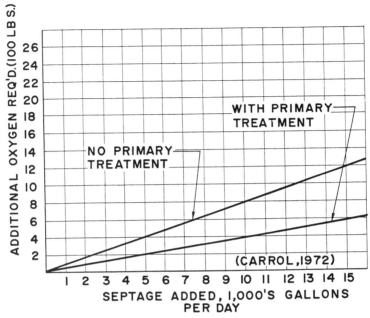

Figure 5. Additional oxygen required for septage addition in activated sludge wastewater treatment plants

higher percentage BOD removal and less sludge production than at the lower sludge age (Figure 6). Wasting must be gradually adjusted with increased loads to obtain a sludge age which produces the optimum between aeration tank efficiency and good settling characteristics. A high sludge age produces a light sludge with poor settling ability but good substrate removal characteristics. The reverse is often true for a very young sludge.

At one plant in New York, septage is bled into the liquid stream inversely proportional to the sewage flow. This procedure takes advantage of a larger excess aeration capacity during lower loading times. Orange County, Florida added septage proportionately with sewage flow rates. Both plants have experienced some operational problems.

Some odor and foaming problems have been reported in aeration systems; however, the odor usually dissipated within 6 to 24 hours [4,21] and foaming was not apparent in all cases. Commercial defoamers, decyl alcohol, heat grids, foam fractional, and aeration tank spray water have been used to reduce foaming.

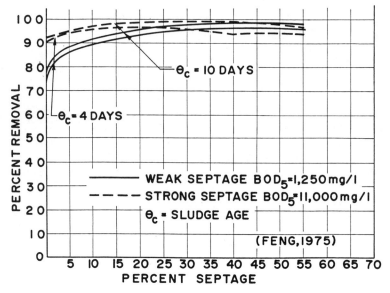

Figure 6. BOD_5 removal from septage-sewage mixtures in batch-activated sludge process.

Attached Growth Systems

Systems that employ attached growth aerobic treatment processes, such as trickling filters and rotating biological contactors, are usually more resistant to upsets from changes in organic or hydraulic loadings, and are suitable for septage treatment.[12,22,23]

In trickling filters, addition recirculation has been shown to adequately dilute septage concentrations and diminish chances of plugging the media. At Huntington, Long Island[12] 30,000-gpd septage is treated at a 1.9-MGD facility. BOD_5 reductions of 85 to 90% have been observed concurrent with total solids and suspended solids reductions of 55 and 85%, respectively.

Rotating biological contactors utilize a long detention time and a continually rotating biological media that is reportedly resistant to upsets. At Ridge, Long Island, a BOD reduction of 90%, COD reduction of 67%, and a TSS reduction of 70% were reported. This installation utilized flow equalization of a low-strength septage. A surface loading of 2 gpd/ft² produced these results.

Aerobic Digestion

An alternative to considering septage as a concentrated wastewater would be to assume septage is the product of an unheated digester, and therefore a sludge.

Many researchers have reported good results in aerobic digestion of septage or septage-sewage sludge mixtures. Jewel[9] reports odors diminished, but time to produce an odor-free sludge varied up to 7 days.

Tilsworth[5] reported a high degree of septage biodegradability at a 10-day aeration time resulting in a BOD reduction of 80% and a VSS reduction of 41%. Chuang,[15] treating anaerobically digested septage with an aerobic digester, reported a 36% VS removal at a 40-day aeration time under a loading of 0.0016 pound VS/ft^3/day on the aerobic digestion. After 22 to 63 days aeration, Howley[14] found a 43% VSS reduction and a 75% COD reduction.

Septage-sewage mixtures were also amenable to treatment. Cushnie[24] showed an average 98% BOD_5 removal from septage-sewage mixtures, ranging from 0 to 20% septage at 6 days aeration. Orange County, Florida adds septage to aerobic digesters at the rate of 5% of the sludge flow, and obtains good reductions at a loading of 0.15 pound VS/ft^3/day. Bend, Oregon obtained good removal adding 13% septage to 87% sludge at a loading of 0.02 pound VS/ft^3/day, utilizing a 15- to 18-day aeration time.[25]

Tilsworth[5] observed α and β gas transfer characteristics for septage, and found that both α, the ratio of gas transfer efficiency compared with tap water, and β, the ratio of O_2 saturation concentration compared with tap water, approached unity after 1 to 2 days aeration. Prior to 1 day, α and β were in the range of 0.4 to 0.6.

Jewel[9] found both dewatering and settleability improved with aeration, but the aeration time required to effect significant improvement varied.

Prior to adding septage to the aerobic digestion process, a check on aeration capacity, toxic metals or chemical accumulations, and increased solids disposal should be investigated. All investigators consistently reported repulsive odors and foaming problems.[5,14,15,24,26]

Studies in high-temperature auto-oxidation of septage are planned[27] and may prove promising as a low-cost, efficient solids destruction technique.

Recommendations, when considering septage addition to aerobic digesters, include: screening, degritting, flow equalization, analysis of excess digestion capacity, and peripheral effects to other processes, such as solids handling. An initial septage addition should be limited to approximately 5% of the existing sludge flow. Additional septage additions should be gradual.

Anaerobic Digestion

Septage is treated in an unheated 20 to 30°C anaerobic digester in Tallahassee, Florida. With an influent septage concentration of 17,700 mg/l total solids, a volatile solids reduction of 56% was reported after an 82-day retention time at a loading of 0.01 pound $VSS/ft^3/day$. Large quantities of grit in the septage required draining and cleaning the open digester after only three years in operation. Leseman and Swanson[28] analyzed volatile acid distribution concentrations in the digester contents. The volatile acid to alkalinity ratio varied from 0.34 to 0.83. The 8-month volatile acid concentration averaged 703 mg/l and ranged from 408 to 1117 mg/l at a consistent pH of 6.0. The progression of volatile acid concentrations in the digester, from two-carbon to five-carbon acids, showed acetic = 276 mg/l, proprionic = 294 mg/l, isobutyric = 14 mg/l, butyric = 49 mg/l, isovaleric = 28 mg/l, and valeric = 42 mg/l. Because this digester had an open cover, gas production could not be monitored. Supernatant from this digester is pumped to the sewage sludge anaerobic digester.

Jewel[9] reported a 45% reduction in VSS from a digester loaded at 0.05 pound $VSS/ft^3/day$, with a 15-day hydraulic retention time. Gas production varied from 4.2 to 7.6 ft^3/lb COD added up to a loading of 0.08 pound $VSS/ft^3/day$, where gas production fell off dramatically, indicating a possible poisoning of the system by a toxic chemical concentration of an unknown source.

Chuang[15] reported a 92% VS removal from a heated anaerobic digester loaded at 0.08 pound $VSS/ft^3/day$ with a 15-day hydraulic retention time. Incoming solids ranged from 0.3 to 8%, and total solids reduction was more than 93%. BOD reductions averaged 75%, from 6100 mg/l influent to 1500 mg/l in the effluent.

Based on his research, Howley[14] recommends a maximum septage addition of 2130 gpd to each 14,500 gallons sewage sludge added/day per million gallons of digester capacity, with a detention time of 30 days, and loaded at 0.08 pound $VSS/ft^3/day$. Good operation of anaerobic digesters requires a limitation on toxic materials.

In single-stage digesters, prior treatment with screening, grit removal and equalization is necessary. Digesters should be cleaned on a regular schedule, such as every two to three years, or as required.

Monitoring digester performance includes long-term evaluation of volatile acid/alkaline ratios and gas production. Mixing is vital in preventing a sour digester from the propagation of point source failure from a septage load containing high volatile acid concentrations.

In systems with multiple tanks, all the above suggestions should be followed. Spreading the septage load to many digesters will reduce septage concentrations. Recycling from the bottom of a secondary digester or from another well-buffered primary digester at a rate of up to 50% of the raw feed per day has been helpful. Control of temperature and mixing should also be adjusted for maximum performance.[29]

Mechanical Dewatering

Islip, Long Island uses a vacuum filter to dewater 100,000 gpd of chemically conditioned septage. A design basis of 6 pounds/hr/ft^2 of surface area was used and appears satisfactory. A lime addition of about 190 pounds/ton of dry solids and 50 gallons/ton of dry solids standard concentration ferric chloride solution are added prior to vacuum filtering.

In another study at Clarkson College, Crowe[30] obtained successful results with vacuum filtration of mixtures of raw septage and digested sludge with up to 20% raw septage by volume. Chemical conditioning with lime, ferric chloride and polymers was required beforehand at chemical dosages typical of domestic sludge. Dewatering characteristics were observed to be similar to those mixtures without septage addition. The filtrate contained only 5 to 10% of the raw septage COD.

Sand Drying Beds

Sand drying has been used to dewater septage, but with varying success. Anaerobically digested septage is reported to require two to three times the drying period compared with digested sludge.[28] After treatment in aerated lagoons and batch aerobic digesters, dewatering simulation studies yielded a septage capillary suction time on the order of 200 seconds, versus about 70 seconds for sewage treatment plant sludges. A lower capillary suction time (CST) can be correlated to a faster dewatering time. CSTs of raw septage usually range from 120-825 seconds, with a mean of 450 seconds.[31] Lime addition of septage prior to sand bed dewatering vastly improved dewatering characteristics. Feige[8] found that an addition of 180 pounds lime/ton dry solids, or 30 pounds/1000 gallons septage based on 40,000 mg/l total solids raised the pH to 11.5, dried to 25% solids in 6 days, and 38% solids in 19 days. An application depth of greater than 8 inches is not recommended because it slows the drying process. The filtrate analysis showed: (1) most heavy metals were tied up in the solids; (2) fecal coliforms were killed effectively; (3) fecal streptococci were more resistant than fecal coliforms; and (4) odors were significantly reduced. Filtrate treatment is necessary, however.[12]

Perrin[31] found other chemicals worked well in modifying the ability of septage to dewater. From a mean initial CST of 450 seconds, septage showed a dewatering ability of 50 seconds after the addition of an average of either 1360 mg/l ferric chloride, 1260 mg/l alum, 1360 mg/l Purifloc C-31, or 2480 mg/l Purifloc C-41.

Perrin also studied the effects of freezing on dewatered samples of septage after treatment in aerated lagoons, or batch aerobic digestion. Freezing lowered the CST from an initial 225 seconds to 42 seconds, an 80% improvement in dewatering time.

If septage is to be placed on sand drying beds, treatment to a consistent CST range of 50-70 seconds is recommended. Further treatment of underdrainage would be required in most cases.

COSTS

Operation and maintenance costs were gathered from existing facilities, operating experience, design reports and estimates.

In all the alternatives investigated, land disposal has the least associated operation and maintenance costs reported, from a low of $1.50 to around $5.00/1000 gallons, exclusive of the cost of the land. Various lagoon systems report cost of treatment between $5.00 and $10.00/1000 gallons. Treatment in sewage treatment plants typically runs about $15.00/1000 gallons, with wide variations in the cost. Composting through the Lebo process costs approximately the same as disposal in wastewater treatment plants. Physical/chemical treatment, such as the Purifax process or chemical stabilization, ranges from similar average costs found in disposal at treatment plants to double or triple that figure.

In a survey of 42 wastewater treatment plants nationwide (Figure 7), only about half charged for septage disposal based on treatment costs.

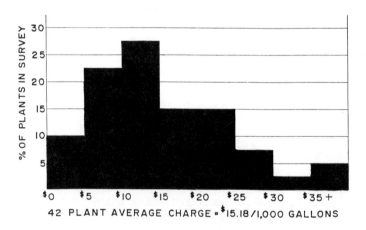

Figure 7. Septage disposal charge at wastewater treatment plants, $/1000 gallons.

Some charge prohibitive rates to avoid septage, while others place a minimal charge on septage to insure against illegal dumping at an unauthorized site. For those plants surveyed, the average charge for septage was

$15.18/1000 gallons. However, an additional 20 to 30 plants contacted either placed no charge on septage disposal, or levied only a yearly fee, most often in the range of $50 to $300/truck.

Many variables affect treatment costs, including local funding requirements; eligibility for state or federal funds; necessity for industrial cost recovery formats; local taxes assessed in lieu of, or to offset treatment plant expenses; level of design; climate; present loading versus design plant capacity; and cost of land. With this in mind, it is easily understood why such a broad range of charges for treatment plant septage disposal was found.

An estimate was performed to determine a reasonable charge a homeowner could expect to pay for having a 1000-gallon septic tank cleaned, assuming no additional work was needed. It was based on a 15-mile haul to the disposal point, 2 hours travel time per load, vehicle depreciation, insurance of $4000 per year, and estimated union wages. Depending on the level of profit, and a disposal cost not exceeding $15, a reasonable charge would be in the range of $40 to $60.

Fees charged to homeowners ranged from a low of $20 to $25/1000 gallons in parts of Long Island to around $100/1000 gallons in areas of New Jersey, Connecticut and Oregon. Rural areas in New England had slightly lower charges, $25 to $40/1000 gallons, while in most areas of the rest of the country, charges were in the range of $40 to $60/1000 gallons. These charges are dependent on the distance from the septic tank to the disposal point (especially pronounced if over 15 miles) and the disposal fee charged.

SUMMARY

Various alternatives for septage disposal have been presented. Good design practices and conscientious operation are necessary to preclude this material from polluting our environment.

The method of choice should depend on a local need evaluation by the design professional, cost-effectiveness of particular solutions, and environmental weighting of impact factors.

For example, while land disposal appears most cost-effective, local constraints in land use, nuisance odors, or poor soil may preclude this option. Similarly, a more expensive option, such as composting, may prove viable because it might meet local requirements for land restrictions, odor prevention, utilization of excess wood waste, and at the same time have a market for its end product.

Sewage treatment plants are among the most frequent acceptors of septage due to their number and location. A properly designed septage handling facility at a sewage treatment plant must be included in any study of alternative treatment schemes.

ACKNOWLEDGMENTS

Information in this paper was obtained under Contract No. 68-03-2231, between the United States Environmental Protection Agency and Rezek, Henry, Meisenheimer and Gende, Inc.

The cooperation of James Kreissl, Project Officer of the U.S. EPA, is greatly appreciated.

REFERENCES

1. "On-Site Domestic Waste Disposal," Draft Report for U.S. EPA by Miller, Inc. (1975).
2. Kolega, J. J. and A. W. Dewey. "Septage Disposal Practice," presented at the National Home Sewage Disposal Symposium in Chicago, Illinois (December 9-10, 1974).
3. Graner, W. F. *Scavenger Waste Disposal Problems,* Report to Suffolk County Department of Health (1969).
4. Goodenow, R. "Study of Processing Septic Tank Pumpings at Brunswick Treatment Plant," *J. Maine Waste Water Control Assoc.* 1:2 (1972).
5. Tilsworth, T. "The Characteristics and Ultimate Disposal of Waste Septic Tank Sludge," Report No. IWR-56, Institute of Water Resources, University of Alaska at Fairbanks, Alaska (November 1974).
6. Kreissl, J. F. U.S. EPA, Cincinnati, Ohio, personal communication.
7. "Septage Treatment/Management Proposal" to Environmental Protection Agency by Maine Municipal Association (May 1976).
8. Feige, W. A. *et al.* "An Alternative Septage Treatment Method: Lime Stabilization/Sand-Bed Dewatering," U.S. EPA Environmental Protection Technology Series (September 1975).
9. Jewel, W. J., J. B. Howley and D. R. Perrin. "Treatability of Septic Tank Sludge," *Water Pollution Control in Low Density Areas,* University Press of New England (1975).
10. Feng, T. H. and H. L. Li. "Combined Treatment of Septage with Municipal Wastewater by Complete Mixing Activated Sludge Process," Report No. Env. E. 50-75-4 for Division of Water Pollution Control, Massachusetts Water Resources Commission (May 1975).
11. Calabro, J. F. "Microbiology of Septage," Ph.D. Thesis, University of Connecticut (1971).
12. "Guidelines for Septage Handling and Disposal," New England Interstate Water Pollution Control Commission, Boston, Massachusetts (1976).
13. "Draft Guidelines for the Design and Operation of Septic and Sewage Treatment Plant Sludge Disposal Facilities," Department of Environmental Conservation, State of New York (undated).

14. Howley, J. B. "Biological Treatment of Septic Tank Sludge," M.S. Thesis, Department of Civil Engineering, University of Vermont (1973).
15. Chuang, F. S. "Treatment of Septic Tank Wastes by an Anaerobic/Aerobic Process," Deeds and Data Supplement, *WPCF Highlights* 13(7):3 (1976).
16. Epstein, F., G. B. Willson, W. D. Burge, D. C. Mullen and N. K. Enkiri. "A Forced Aeration System for Composting Wastewater Sludge," *J. Water Pollution Control Fed.* 48:688 (1976).
17. Epstein, E. and G. B. Willson. "Composting Raw Sludge," Municipal Sludge Management, Proc. Natl's Conf. on Municipal Sludge Treatment, Pittsburgh, Pennsylvania (1974), 123 pp.
18. Tawa, A. Research Assistant, University of Massachusetts, Amherst, personal communication.
19. Carroll, R. G. "Planning Guidelines for Sanitary Wash Facilities," Report to U.S. Department of Agriculture, Forest Service, California Region, H_2M/Hill (January 1972).
20. Feng, T. H. Professor of Civil Engineering, University of Massachusetts, Amherst, personal communication.
21. Feng, T. H. and W.-K. Shieh. "The Stabilization of Septage by High Doses of Chlorine," Report for the Division of Water Pollution Control, Massachusetts Water Resources Commission (June 1975).
22. Design Criteria of Ridge, New York Development Plant, Richard Fanning and Associates (undated).
23. "Town of Wayland, Massachusetts Report on Disposal of Septic Tank Pumpings and Refuse," Weston and Sampson Engineers (November 1969).
24. Cushnie, G. C., Jr. "Septic Tank and Chemical Pumpings Evaluation," M.S. Thesis, Department of Civil Engineering, Florida Technical University (1975).
25. "The Feasibility of Accepting Privy Vault Wastes at the Bend Waste Treatment Plant," prepared for the City of Bend, Oregon by C. & G. Engineers, Salem, Oregon (June 1973).
26. Jewel, W. J., J. B. Howley and D. R. Perrin. "Design Guidelines for Septic Tank Sludge Treatment and Disposal," *Progress in Water Technology* 7:2 (1975).
27. Jewel, W. T. "Waste Organic Recycling Services—Septic Tank Sludge Treatment and Utilization," Proposal to U.S. EPA Region I (June 1974).
28. Leseman, W. and J. Swanson. Water Pollution Control Department, City of Tallahassee, Florida, personal communication.
29. Zickefoose, C. and R. B. J. Hayes. *Anaerobic Sludge Digestion* EPA 430/9-76-001, Municipal Operations Branch, U.S. EPA (February 1976).
30. Crowe, T. L. "Dewatering Septage by Vacuum Filtration," Thesis presented to Clarkson College of Technology (September 1974).
31. Perrin, D. R. "Physical and Chemical Treatment of Septic Tank Sludge," M.S. Thesis, Department of Civil Engineering, University of Vermont (1974).

ADDITIONAL REFERENCES

Bowne, W. C. "An Engineering Study of Septic Tank Content Disposal in Douglas County, Oregon," County Engineers Office (March 1972).

Burkee, D. A. "Handling Septic Tank Wastes," Deeds and Data Supplement, *WPCF Highlights* 7(10):1 (1970).

Cassell, E. A. Department of Civil Engineering, University of Vermont, personal communication.

"Preliminary Engineering Design Report," for Department of Environmental Control Southwest Sewer District No. 3 Water Pollution Control Plant, Consoer Townsend and Associates (April 1971).

"A Study of Septage Handling and Disposal in New England," U.S. EPA Region I, Categorical Programs Division (Draft) (February 1974).

"Papers for Discussion at the Non-Sewered Domestic Waste Disposal Workshop," Office of Research and Development, U.S. EPA Region I, Boston, Massachusetts (April 1975).

Howes, R. Maine Department of Environmental Protection, personal communication.

Isaac, R. A. "Summary of Septic Tank and Cesspool Waste Disposal Methods," Report to Water Resources Commission, Commonwealth of Massachusetts (November 1970).

James, D. W. "Composting for Municipal Sludge Disposal," paper presented at 43rd Annual Pacific Northwest Pollution Control Convention, Seattle (October 1976).

Katsura, Y. "Regional Plant Treats Septic Waste," *Pollution Control* (November 1975).

Kolega, J. J. et al. "Anaerobic-Aerobic Treatment of Septage (Septic Tank Pumpings)," EPA Water Quality Office, Grant No. 17070 DKA Phase I.

Kolega, J. J. "Design Curves of Septage," *Water and Sewage Works* 132 (May 1971).

Kolega, J. J. et al. "Land Disposal of Septage (Septic Tank Pumpings)," *Pollution: Engineering and Scientific Solutions*, E. S. Barrekette, Ed. (New York: Plenum Publishing Co., 1972).

Kolega, J. J. et al. "Septage: Wastes Pumped from Septic Tanks," presented at the Annual Meeting of ASAE, Pullman, Washington (June 27-30, 1971).

Kolega, J. J. et al. "Streamline Septage Receiving Stations," *Water and Wastes Engineering* (July 1971).

Maine Guidelines for Septic Tank Sludge Disposal on the Land, University of Maine at Orono and Maine Soil and Water Conservation Commission, Miscellaneous Report 155 (April 1974).

"Regulations for Septic Tank Sludge Disposal on Land," Maine Department of Environmental Protection (July 1974).

McCallum, R., P. E. "Treat Septic-Tank Wastes Separately," *The American City* (January 1971).

"Draft Report—Task Force on Disposal of Septage and Solid Wastes," Nashua River Task Force, Corps of Engineers (March 1975).

"Septage Survey Report (Draft)," New England Interstate Water Pollution Control Commission (May 8, 1975).

"Appendix D—Septage Disposal Investigation," Report in Tomkins County, New York, by O'Brien and Gere Engineers (undated).

"Septage Disposal Feasibility Study," Report to City and Town of Poughkeepsie, New York, by O'Brien and Gere Engineers (February 1976).

Oliver, J. W. et al. "Heavy Metal Release by Chlorine Oxidation of Sludges," *J. Water Pollution Control Fed.* 47(10) (1975).

Satriana, M. J. *Large Scale Composting*, Noyes Data Corp., Park Ridge, New Jersey (1974).

Smith, S. A. and J. C. Wilson. "Trucked Wastes: More Uniform Approach Needed, Water," *Water and Wastes Engineering* 10(3):49 (1973).

"Pre-Design Report on Septic Tank Sludge Disposal to the City of Portland, Oregon," by Stevens, Thompson & Runyan, Inc. (September 1971).

"Manual of Septic Tank Practice," U.S. Public Health Service 1969, U.S. PHS Publication No. 526.

Vesilind, P. A. *Treatment and Disposal of Wastewater Sludges* (Ann Arbor, Michigan: Ann Arbor Science Publishers, 1975).

Weiss, S. *Sanitary Landfill Technology* (Park Ridge, New Jersey: Noyes Data Corp., 1974).

Weston, R. F. "Preliminary Engineers Report, Septage Disposal Facility Towns of Sudbury and Wayland, Massachusetts," (1975).

"Report on Sludge Disposal for Town of Sudbury, Massachusetts," Whitman and Howard Engineers (December 1969).

Willson, G. B. and J. M. Walker. "Composting Sewage Sludge: How?," *J. Waste Recycling* 14:5 (1973).

"Septage in the Windham Region," Report by the Windham (Connecticut) Regional Planning Agency (December 1975).

"A Study for Disposal of Septic Tank Wastes," prepared for the State of Maine DEP by Wright, Pierce, Barnes and Wyman Engineers, Topsam, Maine (October 1973).

17

COLLECTION ALTERNATIVE: THE PRESSURE SEWER

W. C. Bowne

Director, Special Projects
Oregon Department of Public Works
Douglas County, Oregon 97470

OBJECTIVES

Suppose you were faced with the task of finding a means of alleviating sewerage problems in a small, semirural community which could not economically be served by conventional sewers and had high groundwater and soils of predominantly tight clay. This was the assignment undertaken by the Douglas County Special Projects Division late in 1973.

Had this been a farm area where homes were well separated, there would be comparatively little concern for health hazard. Instead, Glide, Oregon is a community where public contact with effluent from failing drainfields is of indisputable concern. Of the 500 homes involved, 60% of the drainfields show evidence of failure.[1] A building moratorium was imposed, rendering properties unsaleable though land use planning endorsed development of the area.

ALTERNATIVES

Numerous alternatives to the use of conventional drainfield disposal had been reviewed. Some were known to have merit but required either discharge to waters or less restrictive soil and development conditions than those prevalent. Discharge from numerous facilities was not favored nor was it allowed by regulatory authorities. Sites suitable for subsurface alternatives were generally distant from the homes to be served. Surveillance and operational problems were anticipated if a multiplicity of subsurface alternatives were to be used and a sewer system was generally preferred.

Attention was then directed to the engineering studies which had been prepared. Essentially, the conclusion presented in each study was that sewerage costs were beyond the financial capabilities of the area. Thus, conventional sewers were categorized as an infeasible consideration.

Upon further investigation these sewerage studies were found to contain an interesting fact: The sewage treatment plant was estimated to represent only 9% of the project costs.[2] The remaining 91% was for the collection system. Why was the collection system so expensive?

The area is sparsely populated with a resulting long length of sewer between homes. Being mountainous, many obstacles to conventional sewers are present. It appears peculiar that areas such as this are often served by rural water systems, frequently without benefit of grant funds. Why then, does it seem economically impossible to collect the wastewater?

It is known that exclusive of infiltration, wastewater flows would be less than water use because of uses not contributed to the sewer, such as lawn watering. Again, why was sewage collection prohibitively costly when water supply was not? Considerations such as these dictated interest in pressure sewers.

PRELIMINARY RESEARCH

Literary searches performed by the American Society of Civil Engineers, Oregon State University, Water Pollution Control Federation and others resulted in the accumulation of over 40 references. Correspondence was conducted with many of the authors and installations were inspected in Texas, Florida, Indiana, Idaho and Oregon. EPA officials were interviewed in Washington, D.C. and at the research laboratories in Cincinnati, Ohio. The Farmers Home Administration was consulted.

Results of these investigations were encouraging. With strong public acceptance of the engineering report, design is presently underway to provide pressure sewers to serve the Glide, Oregon area.[1]

DESCRIPTION OF PRESSURE SEWERS

The basic elements of one type of pressure sewer are shown in Figure 1. Sewage from the home flows first to the septic tank where floating and settled matter is retained and partially digested. The effluent then flows to a vault where it is pumped into the main and conveyed to a disposal field or plant for treatment.

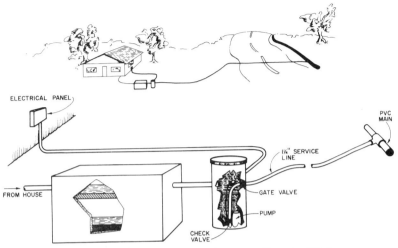

Figure 1. Pressure sewer service connection simplified.

Septic Tank (Interceptor Tank)

Functions of the septic tank are several. Most importantly, it becomes an excellent trap for grit and grease. This benefits pumping substantially and obviates many of the problems associated with the piping system.

A reserve space with a capacity of about one day's sewage flow is also provided within the tank between the normal top of the scum layer and roof of the tank. Should a pump malfunction, sewage flow from the home is not immediately interrupted. A high-level alarm would alert the homeowner whose only inconvenience would be a telephone call to the maintenance office.

A secondary benefit provided by the septic tank is the degree of pretreatment achieved. Studies on the characteristics of septic tank effluent show that reductions of 50 to 60% for BOD and suspended solids may be expected with grease removals of 70 to 90%.[3] Of course, periodic disposal of septage from the tank must be accomplished.

The occasional carry-over of light and filamentous solids may not be detrimental to the pressure system, though this has not yet been fully demonstrated. This might indicate that septic tanks used on pressure sewers could be pumped less frequently than when subsurface disposal is used. The entire tank assembly, including vault and appurtenances, has been termed an "interceptor tank" by Rose.[4] This calls attention to the differences between a septic tank used in conjunction with subsurface disposal and a tank intended for use with pressure sewers where capture of grease, grit, stringy material, and the provision of reserve space are of primary importance.

Commonly, recommendations are to pump septic tanks on about 2-year intervals. However, a study on the subject of septic tank performance by Weibel et al.[5] is interpreted to suggest that longer periods may be reasonable. Experiences by Warren[6] show indications that 500-gallon tanks used in conjunction with pressure sewers may need pumping at intervals of 5 years or even less frequently. Measurements by Schmidt[7] indicate intervals of 10 or more years to be adequate when 1000-gallon tanks are used.

Treatment means used by Warren and Schmidt do not employ subsurface disposal, although they have not suggested whether the intervals would be shortened had this been the case.

In Douglas County, Oregon, the cost of having a tank cleaned is about $40. Though not negligible, the annual cost is seen to be small even if septage were required to be pumped from the tanks fairly often. Treatment and disposal of the septage may be accomplished in a variety of means and should be a matter included in pressure sewer proposals. No elaboration is given here as that subject is beyond the scope of this paper and is covered in other reports.[3,8-10]

Grinder Pumps

Some prefer the use of a grinder pump rather than a septic tank and effluent pump. These grind or cut solids within the sewage to reduce it to a slurry for pumping. However, grinder pumps are more costly than effluent pumps and quite often the entire installation is more expensive than an effluent pumping system, even when accounting for the cost of the septic tank.

Maintenance of a grinder pump is usually reported as being more frequent and more expensive, owing to the grinding function the pump must perform, and the close tolerances common to this kind of pump. Grease is present in grinder systems, sometimes presenting problems with the controls. Also the piping system design is more critical. Grease may accumulate on the crown of the pipe, reducing its capacity and interfering with the action of air release valves. Grease and grit within the piping system dictate that scouring velocities are required,[11] which are sometimes difficult to achieve.[12] Usually, grinder pump vaults are small, providing less reserve space than an interceptor tank. Accordingly, the need for prompt attention in the event of failure is more critical.

For these reasons, effluent pumping has been selected as the preferred practice in the Glide, Oregon installation. In other installations, however,

INDIVIDUAL ONSITE WASTEWATER SYSTEMS 175

it may be important for the sewage to remain aerobic. In these instances, grinder pumps should certainly be considered. Evaluation should be made of the time the sewage remains in the vault and to residence time in the closed pipeline, as the sewage can soon become septic.

Effluent Pumps

One-third horsepower pumps are the most commonly used effluent pumps and usually cost about $150. However, the pumps must be selected based on hydraulic requirements and operating conditions and may vary from one-fourth to two horsepower.[10,11] The matter of simultaneous pumping from a number of installations to a common header (similar to pumping parallel) has been a topic of concern for many. Readers interested in this subject are referred to publications by Battelle[11] and others.[12,13]

Traditionally, engineers have avoided pumping sewage whenever possible. There are important differences to be kept in mind when applying that rationale to pressure sewers:

1. Pressure sewer pumps may easily be removed from the vault and replaced in minutes.
2. Reserve space provides sufficient safety margin to insure uninterrupted service at the home.
3. Grease, grit and stringy material are not present in the pump vault.
4. The pumps are inexpensive.
5. Enough pumps may be involved to justify district employment of a trained and efficient service repairman.

Those who have had limited experience with the pumping of septic tank effluent have a tendency to associate the practice with frustration, when in actuality the reasons for failure are boldly apparent upon careful examination. Typically, they are poorly constructed installations with improperly selected components.

In contrast, the pressure sewer pump installations by Schmidt[7] are an uncluttered and durable design, where maintenance functions can be performed without difficulty and in minutes. All installations are identical, so parts may be exchanged if necessary and maintenance functions simplified. Because of proper maintenance and careful selection of components, the systems designed by Schmidt have operated very successfully during six years of operation.

An installation at Priest Lake, Idaho, having about 500 effluent pumps in operation, experienced problems with 8% of the pumps during the first year of operation.[6] This figure dropped to less than 2% during the second year, and maintenance personnel have anticipated even fewer difficulties in ensuing years.

For the pressure sewer concept to be successful, design excellence is a necessity. Equipment must be selected with great care and installed with the criterion that maintenance functions be made as simple as possible. This becomes a more difficult task than is apparent. Since the advent of pressure sewers is relatively new, suppliers and designers do not have the years of experience on which to rely.

Though most investigators of pressure sewers will be cautious of the need for excessive pump maintenance, well designed installations have proved to be easily maintainable at reasonable expense and a minimum of inconvenience. Readers interested in reviewing pump maintenance data are referred to work by Schmidt,[7] Durtschi,[14] Klaus[15] and others described in some detail in the Glide report[1] wherein an assumption of $50 per year for pump operation and maintenance was adopted.

Service Line

The service line between the pump and main is usually 1¼ inches in size. Installation is easily accomplished with a trencher in contrast to the more difficult installation of conventional sewer laterals.

When sewerage is provided to existing dwellings, homeowners often find the plumbing outlet is oriented to the rear of the house where the septic tank is generally located. Consequently, to connect to a gravity sewer often requires that the house plumbing be reoriented, sometimes at substantial expense.

Mains

Sewer mains are PVC and resemble rural waterline installations. They are sized as dictated by hydraulic design,[11,12] but to describe order of magnitude, Table I may be used. Sizes and costs shown are approximations and should be used for only the most cursory of estimates. To illustrate how widely costs may vary, a recent installation in Texas[16] cost 90¢ per lineal foot for 4-inch pipe as opposed to the $5 per foot estimated for the Glide area.

Lack of extensive data will justifiably cause engineers apprehension with regard to determination of adequate pipe size. At present, there are but a few pressure sewer systems in operation, most sized for a future population. Consequently, pipe sizing may be as yet unrefined, especially with regard to larger systems.

The provision of reserve space within the interceptor tank, and the inherent characteristic that centrifugal pumps can operate at shutoff head periodically, provide a safety factor. Suppose a pump should turn on during a time when the pressure in the main was too great for the pump

Table I. Size and Cost of Pressure Sewer Mains

Number of Homes Served	Size of Main[a] (diameter, inches)	Cost of Main[b] ($/lineal foot)
5	2	3
60	3	4
150	4	5
400	6	7

[a]Pipe sizes have been reported using design flows proposed by Battelle[11] and assuming a velocity of 2.5 fps.
[b]Costs shown are those adopted for the Glide study,[1] where topographic difficulties are more extreme than average. Costs include furnishing and installing the pipe, fittings, valves, bedding, pressure-sustaining devices, road crossings, pipe cleaning, pressure testing, engineering, etc.

to discharge. Then the pump would run without discharging until the pressure in the main lowered. Normally this period would be brief and occasional. Meanwhile, service to the home would be uninterrupted due to the reserve space available. Unless flow from the home continued until the effluent reached the high-level alarm sensor, this condition would not be known. This feature is desirable but not to the point that systems should be undersized with undue reliance on shutoff head operation and the use of reserve space.

Air Entrainment

Desirably, pressure sewers should be oriented so that flow is in the upslope direction;[11] *i.e.,* the outfall should be at a higher elevation than any significant portion of the collection system. Should conditions require that pumping downslope is necessary, large quantities of air may enter the main which can result in hydraulic difficulties. Detrimental effects of air in pipelines are generally known and have been covered in papers by Lescovich[17] and others. The matter of flow in closed conduits on downgrade slopes where two-phase flow may occur is lesser known. However, a paper by Kent[18] describes this condition. To maintain a positive pressure in pipelines several methods have been used; these include the use of vertical stacks or pipe risers and special control valves. Readers are referenced to work by Burton and Nelson,[19] Biggs[16] and Whitsett.[20]

As yet the need for such control on pressure sewer systems awaits further demonstration, but recommendations by Battelle[11] and others[12] suggest that control to avoid two-phase flow and to prevent the entrance of air may well be required.

PRESSURE SEWER VS. CONVENTIONAL SEWERS

In areas where conventional sewers are economically attainable there may be little need to consider alternatives. However, pressure sewers may be feasible when conventional sewers are not. In areas of "difficult" terrain, certain advantages favor the use of pressure sewers and are recounted here for descriptive purposes.

Costs

Under favorable conditions, conventional gravity sewers may be installed at a cost of about $15 per lineal foot. In these cases, and where homes are closely spaced, the conventional sewer is feasible and practical. However, if rock excavation is encountered, prices may rise to $50 or more per lineal foot. Another condition detrimental to the economic installation of gravity sewers would be the existence of high groundwater.

At one installation in Oregon the trench could not be dewatered even when using pumps capable of discharging several hundred gallons per minute. Once the pipe was installed, it suffered many breaks as a result of poor bedding. These breaks, of course, admitted great amounts of infiltration requiring expensive repair to the newly installed sewer.

Cost advantages may dictate the use of pressure sewers under far less extreme conditions than those just mentioned. Where construction within roadways is required, gravity sewer costs might average about $25 per lineal foot. Construction problems may also include the shoring of trench walls, the avoidance of culverts and buried utilities which sometimes require sewer depths to be increased, springs which may be intercepted during trenching necessitating dewatering and many other factors.

When gravity flow in a conventional sewer cannot be continued because of topography or excessive sewer depths, lift stations are required. Though costs vary widely, least expensive lift stations may cost about $15,000. These are infrequently required in areas conducive to gravity sewer collection, but in areas of difficult terrain where pressure sewers would be considered they may be frequently needed. In the Glide, Oregon study,[1] 19 lift stations would have been required had gravity sewers been used. In the pressure sewer proposal this number was reduced to three.

Pressure Sewers Combined with Gravity Sewers

When homes are located at an elevation substantially lower than the route a conventional sewer might follow, the required depth of sewer often becomes great, with resulting high cost. In the conventional sewer option of the Glide study, 48 homes were planned to be served by pressure sewer connections into the gravity main because the homes were at such an elevation with respect to the main that gravity connections were totally infeasible. So pressure sewers can be advantageously applied in conjunction with gravity sewers.

Discussion

Compared with conventional sewers, pressure sewer piping is relatively inexpensive. This allows for sewerage service in extreme topographical conditions or where homes are widely spaced. Also, in using conventional sewers most of the investment must be made in the first stage of development. In contrast, pressure sewers offer a low-cost intrastructure with the cost of the pump and interceptor tank being deferred until the home is built and connected to the main. This consideration becomes significant in slowly developing areas.

Infiltration is common to gravity sewers, often producing wet weather flows of five to ten times that of dry weather. As pressure sewers receive nearly negligible infiltration, a substantial benefit is gained. This must be considered when evaluating these two systems.

After all these factors are taken into account a determination must be made: Will the cost of interceptor tank, pump, etc., and the maintenance required, outweigh the initial cost savings? This question cannot be answered in general; a particular setting must be evaluated. In the Glide study, a 20-year cost-effective evaluation favored pressure sewers by a margin of two to one, as determined by present worth analysis.

While there are many differences to be acknowledged between conventional and pressure sewers, it is presumed the preceding has argued the point for pressure systems sufficiently to acquaint readers with some of the advantages, and perhaps the instances where pressure techniques may be successfully applied.

It is not intended that pressure sewers replace or eliminate the use of gravity sewers. Certainly, in densely developed areas where topographic conditions are conducive to the construction of conventional sewers, an evaluation of the two alternatives may well favor conventional. It is also difficult to evaluate pressure sewers as there are always unknown factors associated with a new concept. Only by experience can the performance of pressure sewers be forecast without some measure of anxiety.

PRESSURE SEWERS VS. ONSITE DISPOSAL

There may be no better means than the use of a septic tank and drainfield for disposal of sewage in appropriate areas. Installed, costs in Oregon average $1550,[21] operation and maintenance requirements are low, and the practice is environmentally sound. Septic tanks and drainfields are normally successful in rural or semirural areas where soils are conducive to subsurface disposal. Alternatives, then, may be suggested when the soils are not suitable for conventional disposal means. As a reasonable cross-section the following choices may be considered: mound systems, sand filters and evapotranspiration.

Mounds

The mound system may be used where soils are not suited to drainfield construction, but only in certain instances described by Otis, Bouma and other researchers at the University of Wisconsin.[22] They are rather large, requiring a suitable site of 2000 to 5000 square feet,[23] which is not always available.

Mound systems are rather expensive with an average installation costing from $3000 to $5500.[23] Carefully executed construction is also required but not as easily accomplished as might be idealized.

Operation and maintenance costs have not been estimated but the system requires the same septic tank and pump as does a pressure sewer system. Mound systems are an endorsed practice and in many areas a good and valid alternative. The choice between a mound system and effluent disposal in another manner will depend on the particular site being evaluated, but pressure sewer components will likely be used in either case.

Sand Filters

Sand filters exist in several designs, notably the intermittent sand filter under study by Otis,[23] Sauer[24] and others, and the recirculating sand filter developed by Hines and Favreau.[25] These systems are reported to treat the waste very effectively, leaving the requirement of disposal. Again, there are options which include (a) disinfection and discharge to receiving waters, or (b) drainfield disposal.

When discharge to waters is employed there is concern as to the reliability of treatment and of disinfection practices. Also, substantial space is required in addition to a septic tank and pump. Costs in Oregon for a single home are reported to be in the order of $3000 to $4000,[26] which includes the drainfield required by the state.

Without the drainfield, costs have been estimated at about $2000.[26] The sand filter alternative to subsurface disposal is thought to have considerable merit and is the system judged most promising by State of Oregon regulatory authorities. However, when serving individual homes, Oregon authorities do not endorse discharge to streams. This is largely due to surveillance problems. Pressure sewers would more likely be considered for groups of homes rather than for single homes. Sand filters may, in some instances, become the treatment method of pressure-collected effluent. When a number of homes can be served, economy of scale can be realized and a responsible agency formed to insure proper operation and maintenance of the single treatment facility. The fact that treatment has been consolidated is of merit and provides a more simple and effective monitoring program.

Evapotranspiration

Evapotranspiration systems are climate-dependent and thus limited in application. Costs may vary widely depending on the particular design employed, but those proposed for experimental use in Oregon are reported to cost from $3000 to $7500 when serving a single home.[27] These systems also require considerable space on the homeowner's property.

Discussion

In recent years considerable progress has been made in developing alternatives to conventional subsurface disposal with results that are highly respected. But the point of this discussion is that each alternative, whether subsurface disposal or conventional sewerage, requires proper application. A large gap exists between those choices, introducing pressure sewers.

PRESSURE SEWER

General Cost

Though there are numerous reasons for use of pressure sewers, economics play a major role. In the Glide study it was estimated that pressure sewers would cost each homeowner $1925 initially and $9.50 per month for management, operation and maintenance.[1] These costs are complete, including the treatment plant, interceptor tank and pump, mains and appurtenances. The capital cost per home is represented in Table II.

Nearly half of the $9.50 charge for operation and maintenance was represented by maintenance of the pump and interceptor tank. Conventional sewers, as previously noted, were estimated to cost homeowners

Table II. Estimated Cost of Pressure Sewer System Per Home
Glide, Oregon

Interceptor tank, pump, etc. (all work on homeowner's property)	$1150
Collection system	475
Treatment plant	300
Total	$1925

about twice as much as determined by present worth analysis, using 6% interest and a 20-year period.

Where obstacles to conventional sewers are even more severe, the cost advantage for the use of pressure sewers widens. An installation in Priest Lake, Idaho, serving 500 homes, was constructed in 1974 at a reported initial cost of one-twelfth that estimated to provide conventional sewers.[14]

Maintenance

It can be argued that the true cost of maintaining pressure sewers will be known only after many years of operating experience. While that is acknowledged, it is also difficult to estimate the cost of maintaining conventional sewers. Historical records from which one would compile statistical cost data have often been gathered from systems with excessive infiltration and inflow and where bypassing has occurred. Assuming that such practices are no longer acceptable, historical maintenance cost records are equally unsuitable for purposes of forecasting.

A similar situation is true regarding maintenance of septic tank-drainfield installations. Often little or no maintenance is given to these systems, but generally their performance has not been satisfactory. One of the factors leading to misconceptions about maintenance required of septic tank-drainfield installations is the lack of adequate records. Where surveys have been conducted, results frequently refute assumptions of satisfactory service.[28]

A basic choice confronts those proposing the use of pressure sewers: Should maintenance of the interceptor tank and pump be performed by the owner or by an established agency? Judging from the maintenance normally provided to septic tanks, owner maintenance is regarded as a risky venture. Also, a valid economic comparison of alternatives can be made only if the systems considered are approximately equal in ability to dispose of sewage without public nuisance or hazard to health. With these thoughts in mind, the Glide study recommended that maintenance be agency-provided. This justifies employment of a qualified service repairman and allows for the more economical purchase of materials and repair. Experience at other pressure sewer projects has indicated that pressure systems when properly managed and maintained will afford a quality of service generally comparable to that obtained from a conventional sewerage system.

Treatment and Disposal

Treatment and disposal may be accomplished by a variety of means. In the Glide, Oregon proposal, a lagoon followed by intermittent sand filters and irrigation disposal is presently under consideration by regulatory authorities. Another alternative is the use of the extended aeration mode of activated sludge treatment with effluent polishing being accomplished by mixed media filtration.

If the number of homes to be served were small, a conventional subsurface drainfield (or alternative) might be used. The pressure concept could offer benefits:

1. The disposal site could be located distant from the homes in a select area.
2. Pressure distribution and dosing principles are often simplified.

In some cases an existing sewer may be close enough that pressure sewer effluent could be discharged into the sewer but where topographic conditions might have rendered the extension of gravity sewers infeasible. In such cases, consideration should be given to three factors: corrosion, odor and toxicity.

Conditions of concern include quantity of septic waste, quantity of receiving sewage, sewer pipe materials and degree of turbulence. These subjects become a far too involved matter for discussion in an introductory paper. Interested readers are encouraged to refer to publications by Pomeroy.[29]

Treatment might be accomplished by a conventional or nearly conventional treatment plant.[3] Though discussion of this aspect is also beyond the purpose of this paper, some differences between pressure sewer waste and conventional sewage should be recognized:

1. Pressure sewer effluent is septic with potential for odors.
2. There is comparatively little history of treating septic tank effluent which would provide basis for design.

Very good results have been experienced by those treating septic tank effluent but in large scale the experience is limited. For the reader's reference, the following are listed: Schmidt,[7] activated sludge; Durtschi,[14] lagoon; Otis and Sauer,[23,24] intermittent sand filter; and Hines and Favreau,[25] recirculating sand filter.

Advantages to the treatment of pressure-collected septic tank effluent are:

1. The waste has been pretreated in a clarifier (interceptor tank). Because of this a grit chamber, bar screen or comminuter would be redundant. The BOD_5 and SS concentrations have been reduced by 50% or more,[3] and little grease is present. Because of the pretreatment provided by the septic tank, simple processes such as sand filters may be used when serving a small number of users.
2. Infiltration and inflow have been nearly eliminated.

In all, the practice of treating pressure-collected septic tank effluent may require further demonstration, but appears promising. An important point to keep in mind is that the treatment and disposal of pressure sewer effluent may be accomplished by any of the methods used in both subsurface practice (or alternatives) and in the treatment of conventional sewage, though modifications may be desired.

SUMMARY AND CONCLUSIONS

Pressure sewers may be used advantageously:

1. When serving individual homes or groups of homes in conjunction with subsurface disposal techniques;
2. To convey wastewater to a receiving sewer; and
3. As an alternative to conventional sewers.

Pressure sewers are particularly adaptable to serving rural or semirural communities where public contact with effluent from failing drainfields presents a substantial health concern.

Benefits are primarily economic, but may include better land use by enabling the development of areas difficult to serve otherwise. Bypasses and overflows common to conventional sewers are obviated owing to negligible infiltration and inflow.

Design requires attention to detail in order to provide a properly functioning and easily maintainable system. Parameters are in the formative stage due to the newness of this concept.

It would seen prudent to encourage the construction of small systems which will acquaint designers with the concepts prior to undertaking more sizable commitments. It is incumbent upon designers of any new system such as this to strive for quality installations. Otherwise, the concept is likely to earn an undeserved poor reputation.

REFERENCES

1. Bowne, W. C. "Glide-Idleyld Park Sewerage Study," Douglas County, Oregon (1975).
2. Cornell, Howland, Hayes & Merryfield. "Proposed Sanitary Sewerage Plan: Glide-Idleyld Park" (January 1972).
3. Vivian, R. "Treatment Study, Septic Tank Effluent and Septage," Stevens, Thompson & Runyan, Inc., Oregon (1975).
4. Rose, C. W. Farmer's Home Administration, Washington, D.C., personal communication.
5. Weibel, S. R., T. W. Bendixen and J. B. Coulter. "Studies on Household Sewage Disposal Systems," Part III, Washington, D.C., U.S. Government Printing Office (1955).
6. Warren, C. Priest Lake Sanitary District, Priest Lake, Idaho, personal communication.
7. Schmidt, H. E. General Development Utilities Company, Miami, Florida, personal communication.
8. Kreissl, J. F. "Septage Analysis," Letter Report, 2/2 (1976).
9. Kolega, J. J. and A. W. Dewey. "Septage Disposal Practices," paper presented at the ASAE Home Sewage Disposal Symposium, Chicago, Illinois (1974).
10. Spohr, G. W. "Municipal Disposal and Treatment of Septic Tank Sludge," *J. Public Works* (December 1974).
11. Flanigan, L. J. and R. A. Cudnik. "State of the Art Review and Considerations for the Design of Pressure Sewer Systems," Battelle Columbus Laboratories, Ohio (1974).
12. Bowne, W. C. "Pressure Sewer Systems," Report presented to Douglas County, Oregon (1974).
13. Environment/One Corporation. *Design Handbook for Low Pressure Sewer Systems* (1973).
14. Durtschi, K. A. Durtschi & Associates, Engineering, Coeur d'Alene, Idaho, personal communication.
15. Klaus, J. G. Klaus Pump & Equipment Company, Portland, Oregon, personal communication.
16. Biggs, J. E. Biggs and Mathews, Inc., Witchita Falls, Texas, personal communication.
17. Lescovich, J. E. "Locating and Sizing Air Release Valves," *J. Amer. Water Works Assoc.* (July 1972).
18. Kent, J. C. "The Entrainment of Air by Water Flowing in Circular Conduits with Downgrade Slopes," Thesis, University of California (1952).
19. Burton, L. H. and D. F. Nelson. "Surge and Air Entrainment in Pipelines," paper presented at conference: Control of Flow in Closed Conduits, Colorado State University (1970).
20. Whitsett, A. M. "Practical Solutions to Air Entrainment Problems," paper presented at conference: Control of Flow in Closed Conduits, Colorado State University (1970).
21. State of Oregon, Department of Environmental Quality. "On-Site Sewage Disposal in Oregon," a status report to the Oregon State Legislature (1976).

22. Otis, R. J., J. Bouma, et al. "Design and Construction Procedures for Mounds," (April 1975).
23. Otis, R. J. University of Wisconsin, personal communication.
24. Sauer, D. K. "Intermittent Sand Filtration of Septic Tank and Aerobic Unit Effluents under Field Conditions," Thesis, University of Wisconsin (1975).
25. Hines, J. and R. E. Favreau. "Recirculating Sand Filter: An Alternative to Traditional Sewage Absorption Systems," paper presented at ASAE symposium, Chicago, Illinois (1974).
26. Ball, H. L. Consulting Engineer, Roseburg, Oregon, personal communication.
27. Ronayne, M. State of Oregon, Department of Environmental Quality, personal communication.
28. Cotteral, J. A. and D. P. Norris. "Septic Tank Systems," *J. San. Eng. Div. ASCE* (1969).
29. Pomeroy, R. D. "Process Design Manual for Sulfide Control in Sanitary Sewerage Systems," U.S. EPA, Technology Transfer Publication (1974).

18

MANAGEMENT GUIDELINES FOR CONVENTIONAL AND ALTERNATIVE ONSITE SEWAGE SYSTEMS— WASHINGTON STATE

Gary Plews

> Program Manager
> Onsite Sewage Disposal
> Department of Social and Health Services
> Olympia, Washington 98504

INTRODUCTION

Traditional patterns of residential development have created sewage disposal problems for health officials. Generally, with newer subdivisions, permanent sewage disposal facilities cannot or have not been installed for a variety of reasons, including the "financial burden placed on the developer." Onsite methods of sewage disposal are therefore used with the understanding that the method is interim and will eventually be replaced by a central collection and treatment facility, when the population base is established to make the permanent facility economically feasible. It's the age-old problem of how to pay for public sewers without the full development to support the costs.

An interesting approach to controlling septic tanks and obtaining sewer service was once suggested by a veteran district engineer:

> "Health departments should not be so concerned about controlling septic tank installations. People should be permitted to do as they wish and put in anything they want. If development is allowed to proceed with no controls, housing starts will be accelerated and septic tank failures will occur more rapidly. Selling the idea of sewers will then be a much easier task."

There is just one small problem with this approach: Who will be responsible for protecting public health during the interim period before sewers are available? This approach also demonstrates the philosophy that public sewers are the only permanent way to handle the problem, so onsite sewage disposal continues to be viewed, for the most part, as interim or temporary.

The origins or sources of development patterns for utilities and the "interim concept" are impossible to trace; but the general ideas very much resemble some principles established in Rome some 2000 years ago. The idea of piping in clean water and transporting wastewater offsite to maximize densities certainly has merit and must be used for highly urbanized existing developments. But, does the concept apply to all developments in all locations? We believe it does not.

TRADITIONAL METHODS

The pattern of "transport offsite" continues to be used and is encouraged by such legislation as PL 92-500, with basin planning, complex facility planning and funding priorities. Health officials are reluctant to change the pattern. There are basic fears that monitoring and enforcement cannot be achieved if permanent onsite systems are used. It is much easier for health officials to "transport" the problem with a collection system and place the responsibility for public health protection with a municipality. The traditional method for utility development and expansion is shown in Figure 1.

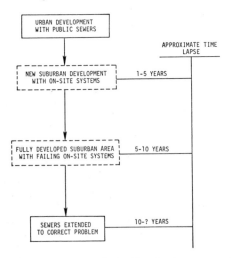

Figure 1. Traditional sequence for utility development and expansion.

Onsite systems have repeatedly received bad press. Failing systems are commonly used as justification for selling the "transport" system. In some cases, a 10% failure rate is considered justification to collect and transport all sewage to a municipal facility. The traditional transport pattern is also assisted by the design standards and maintenance requirements of many regulatory agencies. In some cases, onsite systems are actually designed and maintained to fail because failure brings "super sewers" and transport.

Properly designed and maintained onsite systems can be permanent. The traditional transport concept can be broken if attitudes and management principles are changed. The potential for achieving a high degree of sewage treatment onsite can be realized if certain fundamental changes are made to break the "transport concept." With modern technology, onsite systems can equal or exceed the degree of treatment obtained by most pipe-out methods.

MANAGEMENT CONCEPT

The permanency of the onsite system has been questioned by the engineering profession as well as by health officials. A design life of 30 to 40 years is generally accepted for most municipal systems. Performance of a properly designed and maintained onsite system can match—or may even exceed—the longevity of most municipal systems that have a 30-year design life. Recent longevity studies in Washington, Arizona and Fairfax County, Virginia disclose that standard systems can function from 25 to 30 years. With modern design and with proper maintenance, even longer periods of acceptable performance can be expected.

The real key to changing the traditional patterns is management. If the responsibility for performance of onsite systems is lifted from individual homeowners and placed with a management unit, the potential for having permanent onsite systems is established. Health officials can then deal with a single management system rather than a large number of individual homeowners. Performance of traditional and alternative systems is guaranteed through contracts with the management group. It is responsible for performance and has the authority to install, monitor, maintain and replace systems as needed.

ONSITE SYSTEM MANAGEMENT

The criteria for management are established first in regulation format as follows:

WAC 248-96-070 Onsite System Management. When subdivisions or multiple housing units are designed to have gross densities that exceed 3.5 housing units or 12 people/acre or waste flows of 1200 gallons/acre/day, onsite sewage disposal systems shall not be permitted unless the perpetual maintenance and management of the sewage disposal systems are under the responsibility of an approved management system.

In Washington, public management of sewage systems is preferred; therefore, a developer would be required to obtain management services from a public entity, such as a county, city, sewer district, etc. In the event that no public agency will assume management responsibilities, a special management corporation can be formed. The permanency of the special corporation must, however, be guaranteed by a branch of government through a third-party trust. No totally private management corporation may be formed unless the developer first exhausts all public agency possibilities either directly, or by third-party trust. Figure 2 indicates the general procedure for forming the management system.

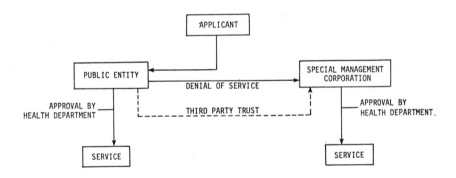

Figure 2. General procedure for forming a management system.

Additional major considerations of a management system are the nature and scope of responsibilities of management and purchasers. In the case of a public agency, responsibilities can be identified and incorporated into local ordinances and rules. Within a special management corporation, however, the responsibility must be identified by contract. Contracts should include, but not be limited to, the following:

1. Agreements to provide maintenance, operation, surveillance, collect fees, keep records, etc.
2. Agreement by purchaser to abide by full conditions as a condition of purchase.

3. A full disclosure procedure.
4. A means of making amendments to contracts by mutual agreement, consistent with original obligations.
5. The right to contract for services.
6. An identification of management's personnel needs and competency levels required.
7. Identification of those portions of system where management shall exercise control.
8. Provision for transfer of management authority to public entity.
9. Provision for purchaser's right to perform certain work.
10. Clear indication that in the event the property is connected to an alternative system, the cost shall be borne by the purchasers.

In addition to the contract items, the structure and criteria for the special management corporation are of key importance to regulatory officials. The corporation must be incorporated under the laws of the state and must have elected officers with a constitution and bylaws. In order to assure proper functioning, the three significant elements for the special management corporation are: (1) a method to insure financial solvency, (2) a method to insure permanency, and (3) a method to implement management functions.

Financial Solvency

There must be financial solvency on a continuous basis. A method of financing must be demonstrated for the construction, maintenance, operation and emergency work of the sewerage systems to the exclusion of whatever other obligations the corporation may assume in other fields. Rates must be set at a level which will provide ample funds for all sewerage operation and maintenance costs, and cover emergencies as they occur.

Permanency

There must be permanency, *i.e.*, the corporation must be continuously in operation so long as there is a need for such management service. There must be built into the organization a provision for transfer of its sewerage responsibilities to a municipal corporation, should such a transfer become feasible.

There must be a municipal corporation (as shown in Figure 1) to which control and operation of the management corporation will pass in trusteeship in the event that no persons are willing to serve as officers of the corporation. In the event that no municipal corporation is able or willing to serve as a trustee, a private organization, acceptable to the regulatory agencies, may serve in this capacity. The municipal corporation or the private organization (third party) shall have the opportunity to review and comment on plans and specifications, perform inspections during construction, and be notified of any future construction or major repairs.

Implementation

During development, the developer must assume complete responsibilities for maintenance and operation of the system. Provisions must be made, with proper controls such as bonding, for transfer of authority to the special management corporation.

The final considerations that should be part of any management system are approvals and flexibility. Any requirements concerning management should be flexible enough to apply to most types of developments. For example, a fourplex living unit on a one-acre tract with good site conditions would probably not require the same degree of management as a 48-unit condominium on a three-acre marginal site. Any guidelines developed for management must, therefore, be flexible, and should generally be approved on a site-specific or case-by-case basis by local government.

SUMMARY

In summary, traditional patterns of residential development have created excessive costs for sewers and have generated health problems. Both situations could be prevented or significantly reduced. Permanent central collection and treatment facilities have generally not been installed initially on new developments for a variety of reasons, including the "financial burden" placed on the developer. Onsite methods are, therefore, designed and maintained as interim or temporary systems to be replaced by a central system when the population base is established to make municipal sewers financially feasible.

The concept of piping clean water in and piping wastewater offsite is presently viewed as permanent and very much resembles an engineering idea developed in Rome some 2000 years ago. This "transport" idea has application to existing, highly urbanized development, but should not be applied to all developments in all locations.

Properly designed and maintained traditional and alternative onsite systems can be a permanent method of sewage disposal. Management and design are the critical factors in establishing permanency. Utilizing modern design criteria will not insure permanency. Establishing an onsite waste management program will reduce health risks, insure permanency, and can reduce the overall consumer cost for sewage disposal.

ACKNOWLEDGMENTS

Within the state of Washington, administrative rules and guidelines are available to establish management systems for onsite sewage disposal. I would like to express my appreciation to Dr. Tim Winneberger, the person most responsible for the concept of onsite waste management. Many of the ideas we used were borrowed from Dr. Winneberger's Georgetown Divide Project in California. The complete guidelines are available from the Washington State Department of Social and Health Services, Office of Environmental Health Programs.

REFERENCES

1. Frink, R. C. and E. D. Hill. "Longevity of Septic Systems in Connecticut Soils," Bulletin 747, The Connecticut Agricultural Experiment Station, New Haven, Connecticut (June 1974).
2. Jewell, W. J. and R. Swan. *Water Pollution Control in Low Density Areas* (Hanover, New Hampshire: The University Press of New England, 1975).
3. United States Public Health Service. *Manual of Septic Tank Practice.* No. 526, U.S. Government Printing Office (Rev. 1967).
4. H.U.D., F. H. A. *Ownership and Organization of Central Water and Sewage Systems,* No. 1300, U.S. Government Printing Office (1966).
5. Prince, R. N. Georgetown Divide Public Utility District, Georgetown, California, personal correspondence (November 1973).
6. Department of Social and Health Services, "Rules and Regulations of the State Board of Health for Onsite Sewage Disposal," State of Washington (June 1974).
7. Vertner, N. C. "A Failure Analysis of On-Site Sewage System in Washington State," Washington State Department of Social and Health Services Report (September 1976).
8. Winneberger, J. T. and W. J. Klock. "Current and Recommended Practices for Subsurface Waste Water Disposal Systems in Arizona," College of Engineering Sciences, Arizona State University, Tempe, Arizona (July 1973).
9. Winneberger, J. T. and W. H. Anderman, Jr. "Public Management of Septic Tank Systems," *J. Environ. Health* 35(2) (September-October 1972).

19
MANAGEMENT GUIDELINES FOR CONVENTIONAL AND ALTERNATIVE ONSITE SEWAGE SYSTEMS— PENNSYLVANIA

William B. Middendorf
    Deputy Secretary for
    Environmental Protection and Regulation
    Pennsylvania Department of Environmental Resources
    Harrisburg, Pennsylvania 17120

    As we move to develop newer methods of onsite sewage disposal and improve on existing systems, we must reflect on the fact that no system is foolproof or failsafe. Although the best design and installation practices are observed, all systems are subject to human misuse and abuse. The suburban and semirural population of today is the progency of the urbanite and suburbanite whose concept of sewage disposal has been limited to the outward appearance of the plumbing fixtures and the payment of monthly and quarterly sewage bills. Without advance warning or forethought, today's suburbanite is faced with the realization that the odoriferous mess in his backyard or his neighbor's backyard is his problem which will not disappear without individual and community action and a large expenditure of funds.
    Government, as a result of this public realization, has reacted by ensuring that future onsite systems are allowed only in certain areas under controlled construction practices. Where such systems have existed in the past but have failed or where further population expansions were contemplated, government's reaction has been the expansion of community collection and treatment facilities. This solution was expressed initially in large complex regional sewer systems that spread from the urban areas to rural America. Said systems were costly and reliant on huge sums of federal construction money. Rising inflationary costs and the decrease in

federal monetary assistance has now led to the awakening that we must rely, in many semirural areas, on smaller, less costly community sewerage systems and, in many instances, at least for the foreseeable future, on individual onlot systems.

The dependence on these types of systems raises problems concerning the proper design, installation and maintenance of the systems.

Historically, government has dealt with these problems by establishing permit systems to ensure proper design and installation. Although the basic permitting system has remained the same in most places over the years, the technical improvements in onlot systems has required a higher degree of competency in personnel representing the permitting authority. Health and environmental agencies and industry have responded to this need by offering sophisticated training programs. For personnel in some permitting authorities such training is mandatory, and in other areas, such as Pennsylvania and Wisconsin, proof of technical competency through certification of permit-issuing personnel for onlot sewage systems is required. Gone are the days when we could rely on simple "perc" tests to determine the acceptance or rejection of onlot sewage disposal systems. We are now in the area of realization that the physical, biological and chemical components of the existing soil or substitute treatment media must be adequately considered before permit issuance. "Out of sight— out of mind" is an old axiom that is now coming back to haunt us and pollute our ground and surface waters. Ensuring proper renovation of sewage effluent is the name of the game whether we are considering stream discharge or onlot disposal for individual or community-wide systems.

When we examine permit issuance as a management concept for sewage disposal systems, it is not difficult to discern that the standards for determining adequate design and installation are often fraught with inconsistencies. In the case of community systems with stream discharges, the problem of inconsistent standards from one regulatory agency to the next has been continually minimized through the action of individual states and the nationwide requirements of the federal government. This factor is not true in the case of standards for onlot subsurface sewage disposal systems. Nonuniform standards persist from one permitting agency to the next. In Pennsylvania as well as in some other states, we have a uniform construction and installation standard which must be adhered to by all local permitting agencies. Nationwide, however, uniformity does not exist and the U.S. Public Health Service *Manual of Septic Tank Practice,* which is not completely up-to-date, has been deviated from in many instances by regulatory agencies either for good technical reasons or for political expediency.

It is understood that design configurations for subsurface disposal will differ depending on flow, soil and geological conditions, but the differences in septic tank design, depth to limiting zone in disposal fields, normal construction material, etc., are normally unnecessary and often founded on nonscientific fact. This has led too often to higher costs for the homeowners in certain areas and hairpulling by representatives of the homebuilding industry and manufacturers of home sewage treatment material and components. Hopefully, the work of the Ten State Committee on Subsurface Sewage Disposal will be successful and, in the near future, will provide us with an up-to-date uniform standard to be considered by all regulatory agencies.

As a management tool, the issuance of permits continues to be abused in many areas. In some areas permit issuance becomes a paper exercise or a revenue collection scheme. It is often difficult to bite the bullet and deny permits where soil, slope or geological conditions preclude the installation of subsurface disposal systems. The regulatory agency has an obligation to alert the general public to the fact that not all sites are suitable for onlot subsurface disposal. In Pennsylvania, we have instituted an extensive public information program utilizing 15-, 30- and 60-second statewide TV spot announcements cautioning potential landowners in unsewered areas to "investigate before they invest." This same theme is expressed in pamphlets placed in the offices of local permitting agencies, lending institutions, real estate agencies and attorneys handling property transactions, etc. The statewide information program is paying dividends in educating the public and taking the "heat" off the state and local enforcement agencies.

To effectively complete the permit issuance program as a management tool it is recommended that permit issuance be integrated with a community sewage planning program. Municipalities should, through a planning process, identify areas which will have to rely for specified periods of time on onlot subsurface systems. Areas to utilize onlot subsurface systems for either interim or long-term usage should be mapped in accordance with existing soil, geological and topographical conditions as well as existing zoning, projected growth patterns and environmental sensitivity. Such a planning system enables local government to realistically plan and control growth and to institute a sewage permitting system which will ensure the planned growth. In addition, local government can also plan institutional configurations to provide small community or home-type sewage disposal arrangements in order to effectively serve planned developments until larger or regional sewer systems can be provided. Such a planning requirement has been instituted in Pennsylvania through the Pennsylvania Sewage Facilities Act. All 2566 municipalities are required to prepare and submit a ten-year plan for sewerage services which outline those items previously mentioned. The plans are approved by the Pennsylvania Department of Environmental Resources (DER). Most municipalities have joined together on a county basis, and the majority of the official sewage service plans are countywide plans. Plans are revised and updated continuously since proposed subdivisions require a revision or update to the master plans. The revisions or updates must also be approved by DER. Permits for onlot subsurface sewage disposal systems to be issued by local government in Pennsylvania or permits for community waste treatment systems to be issued by DER cannot be issued unless said systems are in conformance with the local sewage service plans. The permit management cycle, therefore, is a complete cycle.

In most areas of the country, government has to some degree addressed the problems of controlling the design and installation of onlot systems. Unfortunately, it has not addressed the problems of maintenance of onlot subsurface systems and the need for new innovations in sewage disposal to tackle the horrendous public health problems in small rural villages and settlements that are beyond future regional sewerage systems.

In examining the problem of system maintenance, primary emphasis has been placed on:

1. Requiring that treatment systems employing stream discharges are operated by individuals who have proven their competency through a certification or licensing program;
2. The development and, with the help of industry, the training of personnel who will advise homeowners on the proper care and maintenance of home-type sewage systems;
3. Development of informational and educational material on home system maintenance;
4. The establishment in certain instances of inspection programs for mechanical home-type sewage disposal systems. Many agencies, however, are beginning to contemplate such inspection programs, including a program for anaerobic systems.

We must begin to contemplate other items which affect system maintenance. Serious consideration must also be given by government to mandating low-flow fixtures via local plumbing and housing codes for future homes utilizing onlot subsurface systems. The mandatory submission of routine and perhaps even scheduled pumping of septic tanks as well as the required scheduling and submission of service records for mechanical home sewage disposal systems is also being considered by many local governments. Stronger and more complete training and educational programs are needed for homeowners utilizing home-type disposal systems. Industry can be of great assistance to government in designing and implementing meaningful education programs.

All of these ideas are an outgrowth of the fact that many of our rural and semirural areas will probably never be serviced by a sanitary sewer system, and effective schemes to ensure proper system maintenance must be established.

With this knowledge that rural and semirural areas will not be sewered, attention must now be given to potential institutional arrangements which will resolve existing sewage problems in semicongested areas as well as in certain instances allowing new development with effectively managed smaller community waste disposal systems. Such systems, either those now in effect or in the conceptual stage in various areas of the country, include:

1. The use of individual home sewage holding tanks while sewers are being constructed. Said tanks are then pumped under the supervision of a municipal government or other public entity on a given schedule and the contents disposed of in a nearby sewage treatment plant or on the land under tightly controlled conditions.
2. Individual home and community pre-treatment facilities followed by community subsurface or elevated sand mound disposal systems serving small clusters of homes (two or more) with the disposal site owned and maintained by a public entity. Such systems can employ gravity, vacuum or pressure flows to pre-treatment units.
3. Collection systems followed by pre-treatment and spray irrigation, or in certain areas evapotranspiration, with both the collection and disposal systems owned and managed by a municipality or approved public entity.
4. Public owned and operated collection and treatment facilities with surface water discharge.
5. Any effective combination of the above.

All of these methods have the common management element—a local government or other approved public entity. Such an arrangement assures a reasonable and systematic approach to system maintenance, system design, installation and financial integrity and, most importantly, system responsibility. Private ownership of systems used by two or more households invariably leads to maintenance, financial and property rights problems. For existing problem sewage areas, government should be responsive to any reasonable technical and institutional proposal that will minimize the existing problems. For new development, however, caution on the overuse of such management systems is urged as they greatly stretch any governmental or public entity's ability to effectively control them as well as present serious problems of uncontrolled growth.

In conclusion, therefore, the management guidelines for subsurface sewage disposal systems must take into account methods of ensuring proper system design, installation and maintenance control procedures as well as realistic and applicable institutional arrangements for the provision of sewerage services. Although the majority of the burden for accomplishing these factors rests with government at all levels, it cannot be successfully accomplished without the active participation of the wastewater industry and the understanding and cooperation of the general public.

REPORT ON THE TEN STATE COMMITTEE FOR ONSITE SEWAGE SYSTEMS

Harold D. Baar

>Deputy for Environmental Health
>Bureau of Environmental and Occupational Health
>Michigan Department of Health
>Lansing, Michigan 48909

The Ten State Committee for Onsite Sewage Systems is a committee of the Great Lakes-Upper Mississippi River Board (GLUMRB) of State Sanitary Engineers. The Onsite Sewage Committee was formed in 1975 after a GLUMRB-appointed task force met in January 1975 and determined there is a need for a committee from the 10 states and Ontario to serve as a medium for the exchange of policies, procedures and standards in the individual states, and to serve as a clearing house for the newer development in the area of private waste disposal.

The charge given the committee is: "The examination of the entire problem of individual household sewage disposal problems, including evaluation of research, development of new concepts, development of criteria, consideration of conservation of water use and, finally, development of recommended standards."

The board has set an April 1978 deadline to present a draft of recommended standards for board discussion, with interim reports requested in 1976 and 1977. The committee has held two meetings, in October 1975 and October 1976, to work as a group on proposed standards. Individual committee members have specific responsibility for development of background information and drafts for consideration by the full committee.

The document likely to be created by the committee will no doubt be similar to most state and local health department sanitary codes. The standards for design, construction and maintenance of sewage systems will reflect current technology based on the most recent *documented* research data and experiences of the member states. It is anticipated that widespread use and acceptance of the completed standard will result based on the past experiences we have had with standards developed and published by the GLUMRB. The standards will assume the user already has a basic knowledge of onsite sewage disposal, and will not provide detailed explanations or educational information.

To date the committee has partially completed a portion of its charge by the following:

    A. Preparation of an inventory of each state to determine as accurately as possible the dependence upon onsite sewage disposal as opposed to public or municipal-type sewerage systems.

    B. Compiling in usable form a review of existing standards in each state as relates to onsite sewage disposal, including relationships with local health departments.

    C. Michigan is serving as "clearing house" for the committee by compiling lists of research projects that have been proposed with the intent of avoiding duplication of research efforts. Toward this end, we request that any proposed or recently completed research projects relating to onsite sewage disposal be submitted to either myself or Robert Gurchiek, Bureau of Environmental and Occupational Health, Michigan Department of Public Health, Lansing, Michigan.

    D. A statement concerning evaporation-transpiration systems in the 10-state area has been prepared which states in essence: an onsite sewage disposal system relying solely on evaporation-transpiration is *not* an acceptable method of wastewater disposal. Again, note this statement pertains to the region represented by the committee, as it is based on our collective experiences to date with evapotranspiration systems.

    E. Prepared draft standards for: (1) site evaluation, (2) septic tank design, (3) subsurface absorption systems, and (4) house or building sewer.

    F. Discussed at some length alternative systems and agreed that the following systems needed to be addressed in the standard: mounds, sand filters, privies and compost systems, irrigation, lagoons and aerobic units.

Additional meetings to discuss these alternative systems are considered necessary to meet the April 1978 deadline set by the GLUMRB.

The charge given the committee is a difficult one, complicated by a number of factors the committee has encountered. As you all know, the techniques of onsite sewage disposal are essentially still an *art*, and considerable additional information, experience and research will need to be accumulated before these techniques become a *science*. The verdict on many proposed designs is still out—with the research projects underway unable to provide us with definite answers to questions such as "the ideal mound design," "the use of residential lagoons," "distances contaminants move through various soils," "sizing subsurface absorption fields," "ideal loading rates on various soils," etc.

Also there are wide variances between member state standards, primarily empirical ones, emphasizing the need for reliable research and the need for the development of recommended standards. These differences in member state philosophy seem to result primarily from the local geology differences which influence the political decisions concerning development. For example, a state with rocky terrain or high water table conditions will look with favor upon alternative methods of permitting development as they wish to maintain their population or continue to grow. Another important consideration is the state's attitude on the use of surface water. In Michigan, streams and lakes are heavily used for total body contact; we therefore take every possible measure to protect them for that use. As a result, surface discharges from individual lot systems are not permitted.

Fortunately, the magnitude of the problem involving onsite sewage disposal is being recognized. Additional funds for research are being made available to universities and other organizations capable of conducting reliable research projects. Conferences such as this one help to focus discussion, bringing forth the concerns and stimulating dialog toward addressing the problems. The need for information is critical—our cities continue to lose population while suburban and rural areas continue to grow. The soils suitable for farming are now asked to handle and treat hundreds of gallons of wastewater every day; and, as we all recognize, the heavier soils so ideally suited for farming are not the ideal soils for wastewater disposal.

We believe the committee has made substantial progress toward completion of its task; however, much more remains to be done. Your suggestions and ideas are needed. We urge those of you from the states represented to confer with your representative. Others are invited to communicate with the committee through either Dale Parker as chairman or Bob Gurchiek as secretary. A list of the representatives from each state is provided for your reference.

[Note: Mr. Baar also presented a brief summary of Michigan's Management Plan and Approval Practices for Onsite Sewage Disposal. Refer to pages 55 through 57 of the proceedings of the 1974 National Conference on Individual Onsite Wastewater Systems.]

Great Lakes-Upper Mississippi River Board of State Sanitary Engineers
Onsite Sewage System Committee

Robert Wheatley, Chief
Division of General Sanitation
Illinois Department of Public Health
535 West Jefferson St.
Springfield, Illinois 62761
(217) 782-4674

Durland H. Patterson, Jr.
General Sanitation Section
Division of Sanitary Engineering
Indiana State Board of Health
1330 W. Michigan St.
Indianapolis, Indiana 46206
(317) 633-4393

Robert S. Myers, Sanitarian
Health Engineering Section
Iowa State Department of Health
Lucas State Office Building
Des Moines, Iowa 50319
(515) 281-5344

*Robert Gurchiek, P.E., Chief
Land Subdivision & Planning Section
Division of Community Environmental
  Health
Bureau of Environmental and Occupa-
  tional Health
3500 N. Logan   P.O. Box 30035
Lansing, Michigan 48909
(517) 373-1373

Charles Settle
Public Health Engineer
Division of Environmental Health
Minnesota Department of Health
717 Delaware St., S.E.
Minneapolis, Minnesota 55440
(612) 296-5325

Erwin Gadd, Director
Bureau of Community Sanitation
Missouri Division of Health
P.O. Box 570
Jefferson City, Missouri 65101
(314) 751-4679

Peter Smith, P.E., Chief
Residential Sanitation Section
Bureau of Res. & Rec. Sanitation
New York State Department of Health
Empire State Plaza Tower
Albany, New York 12237
(518) 474-3765

Joe Evans, Unit Head
Home Water Supply & Waste Management
  Unit
Bureau of Environmental Health
Ohio Department of Health
P.O. Box 118   450 E. Town St.
Columbus, Ohio 43216
(615) 466-6434

Glenn Maurer, Chief
Sewage Facilities Section
Bureau of Community Environmental
  Control
Department of Environmental Resources
P.O. Box 2063
Harrisburg, Pennsylvania 17120
(717) 787-9032

**Dale E. Parker, Ph.D.
Soil Scientist
Bureau of Environmental Health
Department of Health and Social Services
P.O. Box 309
Madison, Wisconsin 53701
(608) 266-1704

D.M.C. Saunders
Head, Private Sewage Unit
Pollution Control Branch
Ministry of the Environment
135 St. Clair Avenue West
Suite 100
Toronto, Ontario, Canada M4V 1P5
(416) 965-6967

*Secretary
**Chairman

21

INDIVIDUAL ONSITE WASTEWATER SYSTEM MANAGEMENT IN COLORADO

Edmond B. Pugsley
    Director, Division of Engineering and Sanitation
    Colorado Department of Health
    Denver, Colorado 80220

Most facets of the management of individual wastewater systems in Colorado are controlled by local authorities with some supervision by the State Health Department. A small percentage of such systems is under the complete control of the State Health Department.

Under the *Individual Disposal Systems Act*, the statute which covers individual onsite wastewater systems in Colorado, an individual system is one which has the design capacity to receive 2000 gallons of sewage per day or less, regardless of the manner of treatment and regardless of the means of disposal of the effluent. Under the *Colorado Water Quality Control Act,* the principal water pollution control statute in Colorado, a sewage treatment works is considered to be one with a design capacity to receive more than 2000 gallons of sewage per day. However, in this latter Act, control is had only over those systems which discharge to waters of the state, including groundwater. Because of this, it is interesting to note that apparently there is no statutory control over a system which is designed to receive more than 2000 gallons of sewage per day but does not discharge any effluent to the waters of the state (normally, a sealed lagoon). Another interesting grey area in the statutes is that the latter Act requires site approval by the Water Quality Control Commission for a sewage treatment works designed to serve more than 20 persons, discharging to state waters, with no specifics on the hydraulic capacity. Unfortunately, although the laws were written in the same session of the General Assembly, complete integration and matching were not accomplished.

For the purposes of this paper, we assume that we are talking about what is normally considered an individual sewage treatment system for one or two families. The *Individual Sewage Disposal Systems Act* directs the State Health Department to establish guidelines for rules and regulations which will provide minimum standards for location, construction, performance, installation, alteration, and use of individual sewage disposal systems within the state. The Act further lists the provisions with which such guidelines must comply and, still further, directs that these guidelines shall be the basis for the adoption of detailed rules and regulations by local boards of health.

Local boards of health are required to publish the rules and hold a public hearing on them before final adoption. Then, five days after such adoption, the proposed rules and regulations must be sent to the State Health Department for review. They become effective 45 days after final adoption by the local board of health, unless the State Health Department notifies the local board that the proposed rules and regulations are not in compliance with the statutory requirements. If a local board of health does not adopt rules and regulations, in compliance with the statutory requirement, then the State Board of Health is required to adopt such rules and regulations for that jurisdiction.

It should be noted here that in Colorado there are 13 established health departments, covering 23 of the 63 counties in the state. However, these local health departments do cover approximately 80% of the population of the state. By state law, where there is no organized health department, the board of county commissioners is the local board of health. At the present time, all of the local health departments and the individual counties, with the boards of county commissioners acting as the boards of health, have established rules and regulations for individual systems in compliance with the Act and the guidelines.

It is often said that, in the United States, we have a system that is governed by laws and not by men. However, it must always be remembered that laws are written by men and the Act has some interesting requirements, one of which appears perfectly logical until one realizes that not all areas of the state are near metropolitan areas, and some of them are rather depressed economically. One of the requirements in the administration of the Act is that a final determination as to the suitability of the proposed system be made by a registered sanitarian or a professional engineer. Many counties in Colorado do not have either a registered sanitarian or a professional engineer within their boundaries. This results in considerable inconvenience in certain areas of the state.

Another of the requirements of the Act is that in areas with extremely slow or extremely fast percolation, with high groundwater, high bedrock or steep ground slope, the system must be designed by a professional engineer. Specifics on the above are that where a system is to be built in an area with a percolation rate of less than 1 inch in 60 minutes or faster than 1 inch in 5 minutes, where the maximum seasonal level of groundwater table is less than 4 feet above the bottom of a proposed absorption system, where bedrock is less than 4 feet below the bottom of the absorption system, or where the slope is in excess of 30%—the system must be designed by a professional engineer. In addition, of course, should the individual system discharge into state waters, a National Pollutant Discharge Elimination System discharge permit must be obtained. Because of the complexities of such a permit, we know of no such discharging individual systems in Colorado.

The state's Water Quality Control Commission has the authority and the duty to review the adequacy of local governmental regulations for individual systems in any portion of the state in which the soil, geological conditions, or other factors indicate that unregulated flow from one or more such systems would or might pollute the waters of the state. If the commission, after such a review, finds that local regulations are inadequate, it has the duty to promulgate regulations for the area involved, and shall specify the terms and conditions for individual systems in that area.

The local authorities may charge up to $75 for a permit. Likewise, if the permit is issued under state authority, either where the local authorities should for some reason in the future do away with their permitting group, or whether it is under a permit issued by the State Water Quality Control Commission, the fee is also $75.

The Act prohibits issuance of a building permit for any original construction or remodeling which is not served by proper sewage facilities, *i.e.,* either an individual or a general sewage system.

In order to try to cover all environmental health aspects and to try to avoid duplicate efforts, the State Health Department has a system of memoranda of understanding with local health departments. One of the items in these memoranda is the assignment of responsibility for individual sewage treatment disposal systems. The memoranda, in each case, spell out that the state will adopt guidelines, will review and approve local regulations, and will provide consultation and backup authority for local enforcement as available. The local health departments have the responsibility for conducting the program, except in those instances where difficult areas have been spelled out by the Water Quality Control Commission and the locals have not come through with regulations satisfactory to the Commission. There are a few counties in the state which have not fulfilled this latter obligation. In these relatively few areas, engineers from the Water

Quality Control Division of the State Health Department supervise the individual systems program.

At this point, it seems that a discussion of the effective guidelines is in order. They were passed as temporary emergency guidelines in July 1973, because the *Individual Systems Act* specified that the State Board of Health should adopt them not later than August 1, 1973. As the Act itself was not made into law until June 1973, the staff of the State Health Department had little time to prepare the guidelines as well as they might have liked. The guidelines were readopted after some amendments in February 1974.

Two years' experience with the guidelines has shown that some changes are necessary. One of the most glaring necessities is the matter of allowing for repair of an individual system—normally a typical septic tank with leach field—in a physically restricted area, such as is frequently found in the mountains of Colorado. Minimum distances are specified in the guidelines between various components of an individual system and such things as a cistern, a well or a lake, and so forth. When, after a period of years, a leach field has become clogged and unusable, some homeowners find themselves in a bind because of the lack of space. The present guidelines do not allow for any variance, but variances will have to be brought into the guidelines and into the local boards' regulations. However, there has been a reluctance by various authorities, both local and state, to make changes in these guidelines, pending some further experience with the existing ones.

We have, in Colorado, nearly all generally known individual system types in existence. Our experience with ordinary septic tanks runs the gamut from good to poor. Small aeration units, properly maintained, have performed well but, as is so common, when they are merely put in the ground and forgotten, they are not satisfactory. Transpiration bed removal of effluents—from any system—is being ruled out in some areas because of water rights. Elevated leach fields have worked well in some areas of surface bedrock or high groundwater, or of either too high or too low percolation.

Although the decision as to the applicability of a particular system in a specific location is up to the local authorities, except as specified above in the matter of difficult areas, when a new or unfamiliar system is proposed for a particular installation, the local authorities normally contact the State Health Department for information or advice. If the system is one with which the department personnel are unfamiliar, the department would ask the supplier to make an application to the department for certification of the unit.

The department has an advisory committee on individual systems. It is composed of department personnel, plus several local sanitarians, nominated by the environmental health directors of the state, and named by the State Health Department. This advisory committee reviews the proposed equipment on several bases. If there is an NSF seal, the unit is normally accepted for certification without further consideration. If there is no NSF seal, the committee looks for data comparable to NSF requirements to see if the equipment should be able to meet the criteria of the guidelines. If the committee recommends acceptance and the department accepts that recommendation, the department notifies the local board of health and the supplier of the certification of the equipment. If, on the other hand, the advisory committee recommends nonacceptance, or if the department chooses not to accept a favorable recommendation, the department will notify both the local board of health and the supplier that certification is denied.

The local board of health may still accept the installation but without the certification of the department. It should be noted that no local health department in the state has engineers on its staff.

Should the supplier desire to appeal the administrative decision of the department on a denial, the supplier may request a formal hearing before the State Board of Health. The Board of Health may overrule the department staff or may concur in it. If the board concurs in the denial, then the recourse of the supplier is to the judicial system.

We wish we had better individual systems. As it is, we allow the small systems under what we believe is a balance of treatment performance, groundwater protection and economics. Also, we continually press for central systems, believing that such facilities will be the best whenever feasible.

22

NEW YORK STATE STANDARDS FOR INDIVIDUAL HOUSEHOLD SYSTEMS

Peter J. Smith

    Associate Sanitary Engineer
    Division of Sanitary Engineering
    New York State Department of Health
    Albany, New York 12237

New York State, while pursuing an intense public sewage program for many years, still relies heavily on individual household sewage disposal units. We estimate that as many as 4.0 million New Yorkers or 40% of the population outside the City of New York must depend on such systems to dispose of their household wastewater. In 1975, more than 75,000 field visits were reported by health units involving individual household systems. Additionally, almost 50,000 office conferences were held and over 11,000 sets of plans were reviewed.

STATE STANDARDS

The State Health Department has been publishing recommended standards for household systems since 1930. However, until 1974 these standards just served as guidelines and local health units were permitted to develop local standards. Many of the local standards and policies were customized to meet local conditions. As time went on, many local regulations deteriorated to a point where a solution had to be found to meet every condition, and some local units seemed incapable of preventing the installation system even though early failure was inevitable.

In 1974, Part 75 of the State Health Commissioner's Rules and Regulations was enacted. This identified the State Design Bulletin as the minimum standards. Unfortunately, the design bulletin, which essentially remained unchanged for about 25 years, was outdated and incomplete. Many of the systems and practices commonly used in the 44 local health units were not even mentioned in the bulletin. It was assumed that only new systems were being installed, and that if soil conditions were poor or groundwater came within 4 feet of the ground surface, no development would be permitted. Standards for materials were covered only in a cursory fashion and proper installation, operation and maintenance practices were all but ignored.

To fill this obvious weakness, the State Health Department embarked on the project of producing a revised bulletin. The goal was to publish a more comprehensive document which not only presented effective, reasonable design standards, but would recognize the wide variety of soils, topography and geological formations found in the state. Following the development of an early draft, we requested a thorough review by a select panel of professionals who had demonstrated an interest in the subject and who possessed years of practical field experience. Their contribution to the bulletin was invaluable.

After considerable revision, the next draft was sent to every local health unit, as well as all state departments and agencies with responsibilities for small sewage systems. Additionally, the input from other interested parties, both inside and outside New York State, was solicited. As is so often the case, some responded with constructive and useful comments; most did not.

Finally, last April the bulletin was printed and distribution started. Almost immediately comments and inquiries began to come in. Some complained that certain of the new standards were too strict; others wondered why we had become so lenient. Explanations and clarification were requested. It became apparent that an informational program was in order.

Within a month, we scheduled 11 one-day informational meetings throughout the state. The programs were directed at engineers, sanitarians and technicians working in the individual sewage disposal programs in the county health departments and state district offices. However, these sessions also were attended by representatives of the State Department of Environmental Conservation, U.S. Soil Conservation Service, representatives of local government, designing engineers and even a few installers. Since that time we have met with a variety of interested groups.

The sessions were very productive. The discussions did demonstrate that although we had produced a bulletin which would lead to more uniformity in systems design, we had fallen short of our mark of satisfying the needs of all of our 44 local health units. Those units that wanted more stringent standards presented no problem; these were just minimum standards. Those units, with such poor soil conditions that the average building site in their county would be considered unsuitable, were faced with a dilemma. Another disturbed group were those from areas such as Long Island, with ideal soil conditions who argued that our new standards would increase systems cost by two or three times. How could they justify the cost-benefit to the building industry which was already depressed.

Even after this relatively short time with our new bulletin, we have reached a few conclusions:

1. With some admitted shortcomings, we have developed a manual which should serve to provide uniform good design, installation, operation and maintenance practices in New York State.

2. There is a need to amend our rules and regulations which will give a local health unit which has ideal conditions a method of requesting a general waiver of a requirement of this bulletin from the State Health Commissioner.

3. The rules and regulations should also be amended to permit a local health commissioner or administrator to waive a specific requirement, in an individual case, where such a waiver is justified and is in harmony with the intent of the state standards.

4. Finally, statewide uniformity of standards, while desirable, should only be a step toward developing nationwide standards. This should be our goal.

ADMINISTRATION OF LOCAL PROGRAMS

There is a great deal of diversity in the administration of individual sewage disposal programs which can be found in New York among our local health units. Local units are required to use the state bulletin as minimum standards where there are local programs. However, there is no requirement that there be a local program or how a local program is to be administered. Nevertheless, there are a few common denominators. The state rules and regulations on realty subdivisions establish the bulletin as the minimum standards for housing developments of five lots or more. If a subsurface system is constructed to handle commercial or industrial wastes or is designed for a capacity of more than 1000 gallons per day, a permit from the Department of Environmental Conservation is required.

Aside from this limitation, a county health unit has the option of determining the degree of involvement. The counties supervised by state district offices (there are 10 such offices covering 24 mostly rural counties) normally limit their involvement to plan review of individual systems at realty subdivisions, investigation of nuisances caused by failure of systems, and the providing of advice and guidance on a voluntary basis. County health departments, on the other hand, cover a wide range of activity. Some departments have very sophisticated programs where every installation requires plan review, site inspection, construction permits and the inspection of the completed system. Many tie their program into municipal building inspection programs where building permits are issued only after health department approval, and certificates of occupancy are issued only following certification that satisfactory installation has been completed. A few health departments require that inspections of completed works be made by the designing engineer and he certify compliance with approved plans.

The State Health Department does pay state-aid to the county health departments to cover 50% of the cost of the individual sewage disposal program but has not mandated the level of activity. Reviews of local programs are periodically conducted but are performed primarily to assure that the program is capably staffed and cost-effective. So while we have made in-roads into promoting uniformity of standards, we are still a long way from reaching complete program uniformity.

23
STATE AGENCY MANAGEMENT PLANS AND APPROVAL PRACTICES FOR MAINE

W. Clough Toppan

>Sanitary Engineer
>Department of Human Services
>Division of Health Engineering
>Augusta, Maine 04333

On June 1, 1926 the State of Maine adopted its first plumbing code* which primarily dealt with plumbing within dwellings. The code remained the same from 1926 to 1946, with only very minor modifications. Up until this time very little was written in the code with regard to the disposal of human waste. In 1946 a modification was made requiring a specific length of tile field depending on the type of soil. For example, a sandy soil required 25 feet of land tile per occupant; a silty soil required 60 feet of land tile per occupant. The basis for determining whether the soil was sandy, silty or whatever was based solely on the judgment of the installer. There were no requirements for percolation test or soil identification of any type.

With the advent of more homebuilding in portions of the state the Department of Human Services became increasingly more aware of malfunction problems with subsurface sewage systems due largely to reports from the Farmers Home Administration, the Federal Housing Administration, and state engineers. In 1970 a revision was made to the plumbing code to include a percolation test modeled after the U.S. Public Health

---

*This 1926 code was similar in nearly all respects to the so-called "Hoover Plumbing Code" which was recommended in 1924 for adoption by all the states. It was the outcome of the work of the commission appointed by the U.S. Department of Commerce under Secretary Herbert Hoover.

Service guideline.[1] That modification required percolation tests to be conducted by master plumbers, registered professional engineers and registered land surveyors. As in other states, the State of Maine soon realized that the percolation test had drawbacks because in many instances it did not truly represent the actual soil conditions. Consequently, the state was still faced with malfunctioning sewage disposal systems resulting from poor soils and/or improper installations.

Because of these recurring problems personnel within the Division of Health Engineering of the Department of Human Services conducted an extensive literature survey in 1973. Conclusions drawn from this study showed that the percolation test was not the answer. It was time to drastically modify the Maine Plumbing Code to create a workable system of private sewage disposal that could be relied upon to give a realistic appraisal of the soil conditions and lessen the probability of health and environmental hazards. In July 1974 the State of Maine separated its plumbing code into two parts: Part I[2] pertained to interior plumbing, and Part II[3] to exterior plumbing (private sewage disposal systems). In the new sewage disposal regulations it is now a requirement to have the soil identified by means of an onsite observation hole. This soil evaluation is conducted by registered professional engineers, certified geologists and soil scientists with knowledge and background in soils.

The statutes[4] further require that certification exams be conducted by the Division of Health Engineering and that each town be required to hire a "certified" local plumbing inspector to inspect system installations. As a result of the revised plumbing code, elimination of the percolation test and replacement with a more reliable field test, and the certification program for the local plumbing inspectors in the individual towns, the division began to establish improved management of private sewage disposal throughout the state.

The Division of Health Engineering has divided the counties of the State of Maine into 11 districts. Each district covers an average of 8000 square miles and is supervised by a state-employed sanitarian. Each state sanitarian is called upon by the individual local plumbing inspectors in their respective towns for assistance when needed. As previously mentioned, the legislature required that the municipal officers appoint one or more plumbing inspectors certified by the Division of Health Engineering. The local plumbing inspectors are required to inspect all plumbing within the respective municipality, to insure compliance with state and municipal regulations, and investigate all construction for work covered by those regulations. In addition they were required to change, condemn, and/or reject all work done or being done that did not comply with state regulations. Finally, the local plumbing inspectors are required to keep complete records of all permits, fees and other transactions, and submit an annual report to the Division of Health Engineering.

In the state of Maine there are approximately 485 towns, of which 471 have local plumbing inspectors.

As the result of the certification exam, the local plumbing inspectors regardless of their background are expected to be qualified in three areas: internal plumbing, private sewage disposal, and the laws pertaining to plumbing, subdivision, zoning, etc.

With the advent of the certification program for local plumbing inspectors, changes in the regulations pertaining to site investigations for soil, and the redistricting of state sanitarians, the department felt confident that a workable system had been reached at last. It is difficult to compare this new approach with the former system. In the past the inspections were conducted by master plumbers or general contractors. There was no certification exam, the inspection procedure was poor, and consequently the homeowner was left with a sewage disposal system which was installed with little regard for the existing regulations and the environment.

After the changes of 1973, the homeowner, in order to secure a plumbing permit, had to consult with the local plumbing inspector to determine who was qualified as a site investigator. The homeowner sent the investigator's report to the local plumbing inspector for review. Before issuing a permit the local plumbing inspector had to look at the site or, if knowledgeable of the site, discuss the site with the applicant. Prior to covering the installation, the local plumbing inspector was called and, within 24 hours of the notice, had to visit the installation and inspect it for compliance. Once the certificate of inspection is issued, the application for permit and the certificate of inspection are sent to Augusta and kept on microfilm. Previously the permit procedure was very lax and many times the permits were not submitted to Augusta at all. The new procedure results in good control on the issuance of permits; the department reports that approximately 18,000 permits were issued last year—a significant increase over the 6,000 received in 1972.

In the state of Maine the soils are generally considered poor according to the Soil Conservation Service. Because of the poor Maine soils, systems could not be installed in many areas of the state if the U.S. Public Health Service guidelines were used. For that reason, when the newly formed Part II code was made effective in July 1974, it was modified to be more adaptable to the soils in Maine. An expression once heard was "the soils aren't very good, but they're the best we have." With this philosophy in mind the plumbing code allows sewage systems in the form of beds, trenches, concrete leaching chambers, mounds and special systems. The black and grey wastes can be disposed of in combination or separately.

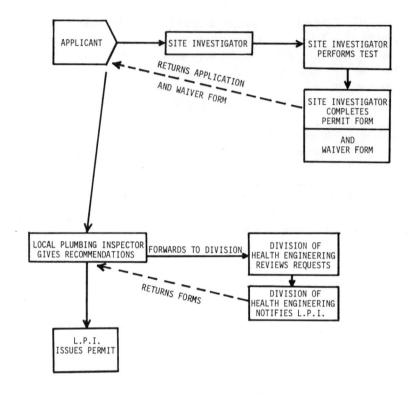

Figure 1. Plumbing permit procedure and waiver procedure.

If separated, with a compost toilet, gas or electric incinerating toilet or recirculating toilet, a 35 to 40% reduction in the size of the system is allowed due to the reduced hydraulic load.

The new plumbing code is oriented toward new construction. With the advent of a minimum lot size of 20,000 square feet, shoreland zoning legislation, implementation of code enforcement officers, etc., the division does not consider new sewage system construction to be a big problem. However, the plumbing code still must address the demands of the existing sewage systems. After examining plumbing codes of other states, it became apparent that the correction of existing malfunctioning systems is many times ignored. Thus a waiver procedure was established to provide for granting waivers to the plumbing code, both at the local level and at the state level, depending on the circumstances. In many instances waivers were granted to eliminate existing overboard surface discharges into lakes, rivers and the ocean. Because of an October 1, 1976 deadline such discharges had to be terminated. Other waivers considered in lots with extremely poor soils fall under the terminology of experimental systems. In many instances septic tanks followed by sand filters followed by leaching beds, raised mound systems, surface application through spray irrigation, and others are utilized. Use of aerobic treatment units followed by subsurface absorption areas is discouraged because the division's engineers cannot justify a reduction in size solely due to the treatment of the waste and because hydraulic loading appears to be the key design factor.

A record of all waivers is kept by the Division of Health Engineering. The department has developed a process of giving the local plumbing inspectors, site evaluators and homeowners authority to waive certain provisions of the plumbing code for repairing existing malfunctioning systems. As more experience is gained, more responsibility will be shifted to local plumbing inspectors and licensed site evaluators.

With the regulation changes, the certification program for the local plumbing inspectors, and the redistricting of the state sanitarians, is the plumbing control program a success? To date it is not possible to conclusively show success; however, considering the results of the numerous inspections throughout the entire state and discussions with plumbing inspectors, the Division of Health Engineering considers that after three years, the program is very successful. Sewage disposal systems are being installed in compliance with the regulations. The result is sewage disposal systems that are harmonious with the environment and provide adequate treatment of the sewage. The division considers this a great step forward.

REFERENCES

1. U.S. Public Health Service. *Manual of Septic Tank Practice*, U.S. Department of Health, Education and Welfare (Washington, D.C.: U.S. Government Printing Office, 1963).
2. State of Maine. State Plumbing Code, Department of Health and Welfare, Augusta, Maine (April 1973).
3. State of Maine. Plumbing Code, Part II, Private Sewage Disposal Regulations (July 1974).
4. State of Maine. Maine Revised Statutes Annoted, Title 30, Section 3222 (1973).

24

INTEGRATED WASTE MANAGEMENT SYSTEMS— ONSITE MIUS APPLICATIONS

W. J. Boegly, Jr. and W. R. Mixon

   Oak Ridge National Laboratory
   Oak Ridge, Tennessee 37830

J. H. Rothenberg

   U.S. Department of Housing and Urban Development
   Washington, D.C. 20410

INTRODUCTION

The MIUS Program is a multiagency program directed toward the development, demonstration, evaluation and ultimate widespread application of a new option for providing utility services to communities. The Department of Housing and Urban Development provides overall direction and sponsorship of the program. Of the numerous government agencies carrying out the program, the Energy Research and Development Administration through the Oak Ridge National Laboratory, the National Aeronautics and Space Administration, the Department of Commerce through the National Bureau of Standards, and the Environmental Protection Agency are major participants.

The MIUS concept utilizes an onsite combined packaged plant, smaller than conventional utility plants, to provide communities of limited size with electricity, space heating and cooling, potable water, and waste treatment and disposal. All major components and subsystems would generally be located within or adjacent to a central equipment building on the development site. Electric and thermal energy would be distributed from the equipment building, using district hot- and chilled-water piping systems, to provide space heating and cooling and domestic water heating; and liquid and solid wastes would be delivered to the MIUS for processing and disposal.

The MIUS concept allows considerable flexibility in order to best serve the needs of any specific community. It could provide all or only part of the utility services; it could be completely independent of or be interconnected with conventional utility systems; it could be owned by the developer or other private entities, a municipality, or a conventional utility company.

A broad objective of MIUS is to provide part of the country's new utility capacity for the next two or three decades with minimal use of critical natural resources, an acceptable cost, and protection of the environment. Hence, an objective of MIUS is to incur a lower total social cost than would occur by use of alternative conventional systems. Also, MIUS is proposed for use only under those circumstances for which its objectives can be met—not as a complete substitute for conventional systems.

Because of the cost of district hot- and chilled-water distribution systems, MIUS is especially applicable to serve multifamily residential developments and associated commercial and public facilities and, to date, the market analysis has concentrated on this sector. Projections of the total demand for new housing to the year 2000 and examination of the size distribution of new HUD-insured multifamily developments in 1970 indicated that about 14 to 36% of total housing requirements might be suitable for MIUS application, depending on the minimum development size for which MIUS is economically feasible. Minimally, we are talking about service to 400,000 housing units per year which would alternately require about 1100 to 2400 Mw of conventional installed generating capacity, depending on climate and the type of building heating and cooling equipment, plus the requirements of associated commercial facilities. If the total of 400,000 new housing units per year was made up of MIUS-served developments, which utilize waste heat for water and space heating and space cooling, the total peak electric demand would be reduced to about 800 to 1000 Mw, depending on climate.

Another objective of MIUS is the provision of service in contingency situations; for example, for use in communities where local restrictions on waste treatment or other utility services prevent construction of housing. The provision of service to isolated communities, both new and existing, which are located too far from municipal-size utility plants for economical transmission of services is also a potential application for MIUS.

A series of evaluation studies on the major components and subsystems of utility technologies applicable to MIUS has been completed as a part of the program.[1-3] Criteria for evaluation included environmental impact, use of fuel energy and other natural resources, economics, and the status of availability. Additional studies included systems analyses of hypothetical MIUS installations in comparison with projected models of conventional utility systems which would otherwise be used. Some of the more important general conclusions from these studies are discussed in the following parts of this paper, in which each of the major utility subsystems of MIUS are examined with respect to the potential for integration with other subsystems and to the advantages resulting from such integration.

ELECTRIC-THERMAL SUBSYSTEM

The MIUS electric-thermal subsystem is a "Total Energy" system in which electricity is generated on the development site. Heat from the water jacket and exhaust gases of engine-generator sets is recovered and used for space heating and cooling and domestic water heating. Based on an evaluation of commercially available technology, the most likely near-term MIUS would utilize gas, diesel, or dual-fuel internal combustion piston engines to drive generators in the 150- to 1000-kw size range.[4]

The heat recovery system might supply 240°F ($\sim$15 psig) steam to single-stage absorption chillers and 200°F water to a district heating system. As shown in the schematic diagram of Figure 1, chilled water would be circulated to provide space cooling by a district piping system. The absorption chiller would be sized and operated to take maximum advantage of available waste heat, and an electrically driven compressive chiller would be used to provide the remaining cooling load. Thermal energy for space heating and domestic hot water heat exchangers in each apartment building could be distributed with a separate district piping system. Auxiliary gas- or oil-fired boilers provide required heat in excess of that available from waste heat. Heat rejection from the system may be by cooling tower or pond.

The system described, and shown in Figure 1, is one concept designed to serve a garden apartment complex, but there are many feasible combinations for which final selection and optimization depend on local parameters. A total energy system inherently integrates the generation of electricity with the provision of residential heating and cooling requirements, and it is the advantages of this integration that will be emphasized in the following discussion.

The primary objective of integrating the electric and thermal subsystems is to increase the beneficial use of fuel energy by the utilization of waste heat. Even the newest of large, conventional power plants releases roughly two-thirds of the energy content of fuels to the environment as thermal pollution and, in addition, about 9% of electricity leaving the plants is

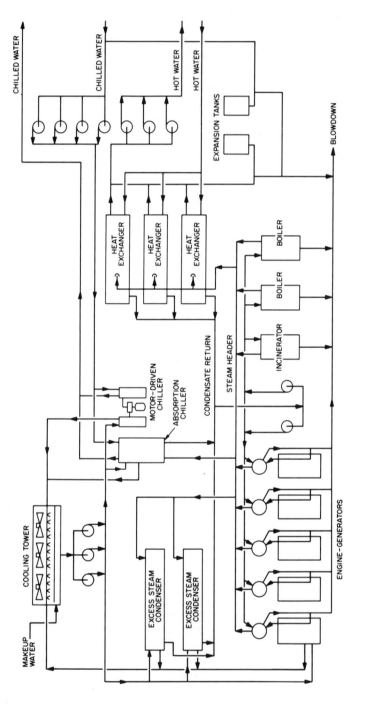

Figure 1. Equipment building schematic flowsheet for the MIUS thermal-electric subsystem.

lost during transmission and distribution. Analytical studies comparing the energy consumption of MIUS with that of alternative conventional systems serving an identical development model clearly show the energy-conserving potential of MIUS.

A typical set of analyses illustrating this point is taken from those completed on a hypothetical, 720-unit garden apartment complex having 60 two-story buildings clustered on 40 acres. The onsite MIUS model employed five 500-kw engine-generator sets (fueled with natural gas), heat recovery and distribution systems, and other components as shown in Figure 1. The alternative conventional models to which MIUS was compared consisted of a central-station electric utility system plus a variety of heating, air conditioning, and water heating equipment within apartment buildings.

Performance of the conventional electric utility model was based on a projected mix of coal-fired and nuclear steam-electric plant additions for the year 1985. The conventional building equipment models consisted of combinations of electric and gas- or oil-burning systems, with all utilizing electricity from the conventional electric utility. The four models considered were: Model C, a conventional district heating and cooling system with central boilers and compressive chillers; Model D, air-to-air heat pumps in each apartment; Model E, large gas-fired boilers in each apartment building plus individual apartment air conditioners; and Model F, electric resistance heaters and air conditioner units in each apartment.

Calculated fuel energy savings realized by use of MIUS instead of conventional models are shown in Figure 2 for five climates. As expected, greater energy savings were realized in colder climates, but significant savings were also found in a year-round mild climate (San Diego) and in a climate with high cooling and low heating demands (Miami).

The energy savings resulting from this example are believed to be conservative. A detailed thermal balance of the MIUS model showed that one-fourth or more of recoverable waste heat was not utilized; partly because of the noncoincidence of the time that waste heat was available and the time that heat was needed, and partly because of the simple garden apartment model used. There are many possible, and perhaps more realistic, consumer models and techniques to optimize the total MIUS-consumer complex which could effect an increased utilization of waste heat. Success of an integrated total-systems approach, such as the MIUS concept, requires careful planning and coordination and a concern for life-cycle costs. A definite advantage of these requirements is to provide the mechanisms and incentives to realize the fullest potential for conservation of energy and other resources.

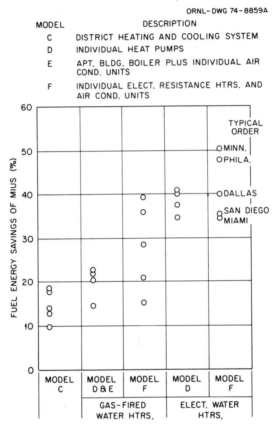

Figure 2. Fuel energy savings of MIUS over various conventional HVAC models serving 720 garden apartments in five cities. (Based on 1985 mix of new conventional central station plants with 33% generating efficiency.)

Utilization of MIUS would effect a significant savings in the consumption of fuel energy, but the impact on a particular fuel resource depends on the type of fuel used in MIUS and in the alternative conventional systems used for comparisons. New conventional utilities were projected to use coal and nuclear fuel to generate electricity and some combination of gas, oil or all-electric systems within buildings. On this basis, MIUS, using gas- or oil-fueled systems, would consume relatively more gas (or oil) than conventional utilities. Of course, if some mechanism should exist by which gas (or oil) to be used in MIUS could be reallocated from less efficient users of the same fuel, then the use of MIUS would result in a savings of that resource.

In order to avoid possible limitations in the use of MIUS due to shortages of oil and natural gas, and to provide a more positive contribution to the conservation of these premium fuels, the Department of Housing and Urban Development and the Department of Interior are jointly sponsoring the development of a coal-burning MIUS. This concept, under development at the Oak Ridge National Laboratory, utilizes a coal-fired, fluidized-bed combustion chamber coupled to a closed-cycle gas turbine to generate electricity. As in current MIUS concepts, waste heat would be recovered and utilized for heating and cooling requirements.

Increased utilization of fuel energy effects a corresponding reduction in heat released to the environment. In the MIUS models used for systems analyses, the overall thermal efficiency of engine-generator sets was between 60 and 70% for annual operation in all climates except Dallas; but more optimum installations, or an improved match between electric and thermal loads, could increase this value to 80%. With respect to thermal efficiencies, a MIUS operating at 70% rejects to the environment less than half the waste heat of a large modern steam-electric plant operating at 36%. In addition, heat rejected by MIUS would generally be released with exhaust gases at temperatures not exceeding about 300°F and not to natural waters or evaporative cooling systems. Cooling towers would normally be required for use with absorption and compressive chillers, but operation would be limited to the cooling season.

The MIUS models used for analyses were designed to include excess capacity and a multiplicity of components in order to provide services with a reliability at least equal to that of alternative conventional utilities. As an example, the electrical subsystem of one model[2] was designed to meet peak winter electricity demand with only two of five generators in operation. Three generators met peak summer demand and shutdown of one of these would cause a loss of only part of the space cooling—that provided by compressive chillers. The design allowed at least one, and at some times two, outages while one was down for planned maintenance with no loss of service. Reliability was similarly included in the design of auxiliary boilers, pumps and other critical components. Operation of only one generator with auxiliary boilers could provide space heating and electricity for critical loads. Other options could include connections with a conventional electric utility grid, either as a backup or as an integral part of the system. Thus, the probability of a complete loss of all utility services could be quite low.

Advantages of integration of the electric and thermal subsystems are summarized as follows:

- Beneficial use of fuel energy is significantly increased and energy consumption is correspondingly reduced.

- Thermal energy rejection to the environment, especially to surface water bodies, is reduced. Cooling water withdrawal and consumption is reduced.
- The reliability of space heating and electric service can be improved.
- In many situations, waste heat would be economically available for use in other subsystems and in special applications such as ice melting, heating pools, commercial process heat, etc.

SOLID WASTE SUBSYSTEM

No single method of collection or disposal of solid waste is applicable to every MIUS configuration or to all geographic locations. The feasibility of various methods of collection depends on the design and layout of the dwelling units and the type of disposal operation selected. Selection of the method of disposal depends on the extent of desired integration with other subsystems, such as the need for heat which incineration could provide.

The currently available collection options include such operations as compacting or bailing before pickup; manual collection with vehicular transfer; and the more innovative vacuum, slurry or pneumatic collection systems. With proper consideration for the method of disposal, any of the collection methods could be used in a housing development whether services are provided by MIUS or conventional utilities. However, the use of onsite waste processing would be well suited to the application of innovative automatic collection systems. Such systems would replace manual handling, local storage, littering and vehicular traffic which are estimated to account for about 80% of the total cost of conventional waste collection and disposal. The integration of planning, operation and ownership of all utility systems, and possible transfer of responsibility to the developer or residents, should provide considerably more incentive to optimize collection methods for each development.

The population of the MIUS-served development significantly affects the methods of solid waste disposal to be considered. Developments in the size range of 300 to 3000 dwelling units would require disposal of about 14 to 140 tons per week of solid waste, respectively.[5] Disposal systems, such as incineration, operating 5 days per week and 7 hours per day would be required to process about 0.4 to 4 tons per hour over that range of development size. Capacities in this range are considered small in the field of solid waste disposal; as a result, very little development work has been done and there is little field operating experience on systems of MIUS's size.

Disposal methods which produce effluents usable in other MIUS subsystems include incineration and pyrolysis. Pyrolysis reduces waste volume and produces gases which may be of use as fuel, but pyrolysis plants are not now commercially available in sizes applicable to MIUS.

The recovery and utilization of waste heat inherent in MIUS applications together with the goal of energy conservation make use of incineration with heat recovery a very attractive option. An evaluation completed in 1973 indicated that incinerators which could be incorporated in a MIUS were currently available (with and without heat recovery), that new and unique designs were under development, that emissions could represent only a small contribution to air pollution, and that the economics of employing incineration with heat recovery might be acceptable.[1,2,5] Further development, demonstration and operating experience are necessary to make firm conclusions.

Systems analyses comparing fuel requirements of a MIUS which includes incineration with heat recovery to that of a MIUS without incineration are complex and depend on many assumptions which affect the amount of waste heat available from generation of electricity, the amount of auxiliary incinerator fuel required, the amount of usable heat from incineration, and the coincidence of the time when recovered heat is available to that when it is needed. Dwelling and MIUS heat balance calculations were completed using hourly weather data with a particular MIUS application serving 720 garden apartments, with and without solid waste incineration. With no heat storage capability and no auxiliary fuel required for incineration, the results indicated that use of incineration with heat recovery would reduce total fuel consumption of the MIUS model by about 2% with San Diego and Miami weather, 3% with Dallas weather, 4% with Philadelphia weather, and 5% with Minneapolis weather.

Most of the solid waste currently produced is disposed of by the use of open dumps, burning, and land disposal practices which do not meet EPA or state criteria for approved operations. These methods cause health hazards and extensive air, water and soil pollution. Because of the recognized need to upgrade existing facilities and the high standards for new facilities, the waste management problems facing communities and the cost of disposal are increasing rapidly. Types of incinerators considered for MIUS application would reduce residential waste to a sterile residue with a weight reduction of ~ 80% and a volume reduction of ~ 80 to 90%. Residue from incineration, and bulky noncombustible items, would still require ultimate disposal in landfills; but the significant reduction in land area requirements, impact on the land used, and vehicular traffic to the landfill are important advantages of incineration. Incentives necessary to realize these advantages in the near future should be insignificantly increased by the integration of incineration with other MIUS subsystems and the improvement in economic factors from beneficial use of waste heat. With currently available technology, the most likely solid waste disposal method which contributes to MIUS objectives appears to be incineration with waste heat recovery.

Advantages of integrating the solid waste subsystem with other MIUS subsystems are summarized as follows:

- Use of innovative waste collection methods, with reduced environmental and community impacts, is more likely.
- The cost and fuel consumption for solid waste transportation is reduced.
- Land area requirements for ultimate disposal are reduced compared to conventional landfill disposal.
- Health hazards and environmental impact associated with landfill operations are reduced or eliminated.
- The contribution of recovered waste heat reduces fuel consumption.
- Use of thermal storage or selective burn periods could increase the beneficial use of recoverable heat by providing thermal energy at times when insufficient waste heat is supplied by other MIUS subsystems.

LIQUID WASTE COLLECTION AND DISPOSAL

Liquid waste collection, treatment and disposal can be performed within the utility framework of the MIUS using currently available technologies.[6] The small size and self-contained nature of the MIUS create special situations with respect to the application of existing and future technologies and the integration with other subsystems in the MIUS. In general, conventional waste treatment systems are not large consumers of electricity or waste heat; however, new concepts such as distillation, reverse osmosis, or elevated temperature biological processes can consume electricity or waste heat. The major advantage of integration is expected to be the use of treated effluents for irrigation, fire protection, industrial uses or cooling water.

Pressure or vacuum sewers for new installations might have advantages over conventional gravity-flow sewerage systems. Because such sewers do not have to follow the hydraulic grade lines required for gravity sewers, excavation could be reduced.[7,8] In addition, capital costs might be reduced; however, operating and maintenance costs would be higher than those for gravity sewers. The vacuum sewer, used in conjunction with a specially designed vacuum toilet, reduces water use in dwelling units, and other water-conserving plumbing fixtures are also available. Integrated management and the importance of life-cycle costs to a MIUS owner may provide incentives necessary for use of water conserving systems.

Small, factory-built sewage treatment plants that could be used in a MIUS are commercially available from more than 50 manufacturers.[6] The small size of a MIUS (less than 0.5 million gallons per day) makes the use of a plant that could be prefabricated and shipped to the installation site attractive. At the time of an evaluation of available technologies applicable to MIUS, most package plants used biological treatment processes, and only about five physical-chemical treatment units were commercially available. However, liquid waste treatment is currently undergoing rapid change. New systems such as physical-chemical treatment and biological treatment supplemented by physical or chemical treatment have been developed and are being used to effect high removal of suspended solids, nutrients and trace organics.[9] These systems, many of which are available in package plants, are designed to provide effluents of high enough quality to be reused for fire protection, irrigation, cooling, recreational purposes, etc. Direct recycling of effluents for use as potable (drinking) water is also feasible but awaits further study for the resolution of such problems as virus control, reliability of plant operation, and monitoring methods.

Addition of more unit processes to the biological treatment system, or the use of a physical-chemical plant, is more expensive than the simple biological plant. Thus, the choice of package plant type will depend on many site-specific conditions such as the intended reuse of treated effluent and the quality, use and size of receiving water-bodies.

The 720-unit garden apartment model required about 200,000 gpd of domestic potable water for its inhabitants resulting in the production of about 200,000 gpd of wastewater to be treated. In addition, cooling towers in the electric-thermal subsystem were estimated[2] to require about 4.6×10^6 gal per year of makeup water. Thus, use of treated liquid waste as cooling water would decrease water withdrawal by about 6%. Potable water treatment plant capacity and water withdrawal rates can be reduced accordingly. Additional savings in water withdrawal and water system capacity would be realized by using treated liquid waste for lawn watering, but amounts involved are site- and climate-dependent.

Whether it is more economical to treat liquid waste in an individual MIUS or to install an interceptor sewer to the nearest available waste treatment plant will depend on the size of the MIUS, the capacity of the existing plant, the geologic conditions and topography affecting the cost of an interceptor sewer, and the projected growth patterns for buildings and utilities in the urban area. In addition, it might be possible for a number of MIUS complexes to use common collection or treatment facilities. Thus, each MIUS and urban growth pattern projection must be evaluated separately.

Advantages of integrating the liquid waste subsystem with the other MIUS subsystems are as follows:

- With high-density housing in a MIUS, it may be feasible to provide innovative collection systems such as vacuum or pumped sewers.
- Onsite liquid waste treatment systems eliminate the need for offsite interceptor sewers to serve the development.
- In areas where existing treatment plants are overloaded, integration of liquid waste treatment into MIUS allows provision of waste treatment in phase with community development.
- The liquid waste subsystem may be able to use waste heat from the electrical-thermal subsystem to improve treatment efficiency.
- Sludges produced during liquid waste treatment can be incinerated and any recovered heat returned to the overall system.
- Proper design and operation of the waste subsystem can provide effluents that can be reused for fire protection, process uses, irrigation, lawn watering, or cooling purposes, thus reducing water withdrawals from natural or potable sources.

POTABLE WATER TREATMENT

MIUS may provide potable water for drinking, bathing and cooking with an onsite water treatment plant. The source of water may be groundwater or a surface water supply. If groundwater is used, the treatment may consist solely of disinfection, unless impurities such as iron or hardness must be removed. In the case of surface water sources, treatment may consist of coagulation, sedimentation, filtration and disinfection. A number of manufacturers produce package plants which are capable of being used in MIUS.[10] In certain parts of the United States it may be necessary to provide desalination to reduce dissolved solids, and equipment in the size range required by MIUS is currently available. In general, integration with other MIUS subsystems will be at a minimum if groundwater is used as the water source, and a maximum if desalination technology is required.

Advantages of integrating the MIUS potable water subsystem by including onsite treatment facilities include:

- No dependence on outside water sources.
- The need for obtaining additional land for constructing costly transmission lines is eliminated.
- The need for increased capacity at existing overloaded treatment facilities is eliminated.

- Small MIUS size increases the possibility of using groundwater sources which usually require only minimal treatment for potable use.

COMMON ADVANTAGES

Several advantages of subsystem integration are common to all subsystems or result generally from use of the MIUS concept. MIUS subsystems may share operating personnel and land, buildings, and other site improvements associated with utility systems. A MIUS using currently available technology can provide reliable service completely independent of conventional utilities. Thus, utilities can be installed in phase with requirements of developing communities and the need for offsite transmission of services is eliminated. Because many new utility components are first proved with units small enough for direct application to MIUS, the potential to rapidly incorporate advanced technologies in MIUS is considered at least equal, and perhaps better than in conventional systems.

Results of evaluations indicate that MIUS can meet many of the program objectives while providing services at about the same cost as alternative conventional utilities.[1,2] While MIUS is not proposed as a competitor or a complete substitute for conventional utilities, there are believed to be a significant number of applications for which MIUS would be of public benefit. It is expected that municipalities and utility companies, as well as private developers, would consider this option.

ACKNOWLEDGMENT

Research sponsored by the U.S. Department of Housing and Urban Development under Union Carbide Corporation's contract with the Energy Research and Development Administration.

REFERENCES

1. Miller, A. J., G. Samuels, W. J. Boegly, Jr., and L. Breitstein. "Technology Evaluation for MIUS," in *8th Intersociety Energy Conversion Engineering Conference Proceedings,* University of Pennsylvania, Philadelphia, Pennsylvania (August 13-16, 1973).
2. "Draft Environmental Statement—Application of Modular Integrated Utilities Systems Technology," HUD-PDR-EIS-75-1F (October 1975).
3. Oak Ridge National Laboratory. "Energy Division Annual Progress Report," ORNL-5030 (December 31, 1974).
4. Samuels, G. and J. T. Meador. "MIUS Technology Evaluation—Prime Movers," ORNL-HUD-MIUS-11 (April 1974).

5. Boegly, W. J., Jr. et al. "MIUS Technology Evaluation—Solid Waste Collection and Disposal," ORNL-HUD-MIUS-9 (September 1973).
6. Boegly, W. J., Jr. et al. "MIUS Technology Evaluation—Collection, Treatment and Disposal of Liquid Wastes," ORNL-HUD-MIUS-16 (December 1974).
7. Carcich, I. G., L. J. Hetling and R. P. Farrell. "A Pressure Sewer Demonstration," Report EPA-R2-72-091, Environmental Protection Agency (November 1972).
8. Mekosh, G. and D. Ramos. "Pressure Sewer Demonstration at the Borough of Phoenixville, Pennsylvania," Report EPA-R2-73-270 Environmental Protection Agency (July 1973).
9. Kugelman, I. J. "Status of Advanced Waste Treatment," report presented to the Long Island Marine Resources Council, Hauppauge, Long Island, New York (1971).
10. Compere, A. L., W. L. Griffith, W. J. Boegly, Jr., I. Spiewak, D. G. Thomas and S. A. Reed. "MIUS Technology Evaluation—Water Supply and Treatment," ORNL-HUD-MIUS-21 (April 1976).

25

THE BOYD COUNTY DEMONSTRATION PROJECT—
A SYSTEM APPROACH TO INDIVIDUAL RURAL
SANITATION (AN UPDATE)

Lawrence E. Waldorf

>Senior Program Analyst
>Appalachian Regional Commission
>Washington, D.C. 20235

INTRODUCTION

Two years ago at the first NSF Onsite Wastewater Conference, the Appalachian Regional Commission (ARC) was honored to present a discussion of a new project being developed to provide rural families with sanitary facilities in areas where the cost of conventional collection and treatment have proved to be prohibitive. The project, known as the Boyd County Demonstration Project, initiated the "System Approach to Individual Home Treatment." This paper is a candid discussion of that project. It will provide background for those who may not be familiar with the project; discuss some of the more interesting installations in the project; talk about lessons learned from the project (which is the purpose of any demonstration), and project a future outlook for such projects, perhaps even offering a few pieces of unsolicited advice.

BACKGROUND

Traditionally there have been two choices, once it is agreed that rural sanitation is a problem which should be addressed. At one end of the spectrum has been municipal collection and treatment facilities; at the other extreme, individually owned and maintained septic tanks, outhouses or some other "devices." Between the two extremes there has been a great void; and, when either the cost of municipal facilities was too high or septic tanks would not function, nothing was done. In rural Appalachia, not to mention the rest of the United States, for hundreds of thousands of families, the cost of municipal collection and treatment is not economically feasible, even with 75% grant money from EPA. Because of the combination of low population density and severe topographical problems, the cost of providing municipal collection and treatment is now regularly between $5,000 and $10,000 per house. For example, in West Virginia's Hepzibah Public Service District, a system designed to serve 150 homes had an estimated cost of $1.2 million or $8,000 per house. In Garrett County, Maryland, a system built with ARC funds cost $8,500 per house. Another system in the same area cost $8,700 per house. The result of these high construction costs has been high user fees such as that experienced by a small community in Monroe County, Pennsylvania. Here, in order to finance a municipal collection and treatment system, even with the assistance of both EPA and ARC funds, a tap-on fee of $500 was charged, and a service charge averaging approximately $20 per month assessed. These are actual projects in our records at the commission, and it was this kind of project which prompted ARC to try to find some alternatives which would help to fill the gap between municipal treatment and individually maintained septic tanks.

There are still 2,700,000 homes in Appalachia alone which do not have access to public sanitary facilities. One of the primary reasons for this is that the option of high-cost municipal collection and treatment facilities, with their resultant excessive monthly charges, is not suited to the needs of the people in many rural areas. As Senator Randolph has stated before in the U.S. Senate,[1] 19,500,000 households across America are not served by public sanitary facilities, and these families must provide for themselves some method of home disposal for the nearly 3 billion gallons of domestic sewage which they generate daily. These conditions exist despite the appropriation by Congress of tens of billions of dollars for the construction of sanitary facilities.

The "System Approach to Individual Home Treatment" is an attempt by ARC to develop a tool which can be used to fill the gap between municipal treatment and collection, and individually owned and maintained septic tanks. It is not *the* answer to the problems of rural sanitation. No one has *the* answer, because the problems are so diverse that only a serious, ongoing program to improve and develop new tools which can become parts of a total answer is a realistic solution. What is needed is an inventory of tools between the two extremes, from which the rural sanitary engineer can draw those which are best suited to his particular situation. The system approach is an attempt to develop one of those tools, to be one component of that inventory.

The Boyd County system approach is based on two important assumptions:

1. The average homeowner either cannot, or will not and probably should not, properly assume the maintenance of his own sanitation device.
2. Rural families are entitled to the same quality of public service as those living in more urbanized areas.

Our project, therefore, is based on the premise that rural sanitation must be treated as a public utility, *i.e.*, all equipment involved must be owned, operated and maintained by a public body—in this case, a public sanitary district. For purposes of system maintenance and eligibility for federal grant funds, this concept is essential.

PROJECT DESCRIPTION

The project area is located approximately five miles from the Huntington Airport in Kentucky. The area has the typical characteristics of low population density and rough topography found throughout Appalachia. There are about 60 families living within the boundaries of the sanitary district. This project serves 47 of these families and includes 36 individual home aeration treatment plants, and 2 multifamily aeration plants serving 11 families. One of the goals of the project was to build into the system for demonstration purposes as many variables as possible with respect to equipment installation. The aeration equipment being used in Boyd County is manufactured by Multi-Flo, Cromaglass, Flygt, Bi-A-Robi, Jet, and Nayadic. Most of the conference attendees are probably familiar with this equipment, because each of these manufacturers has received the NSF seal.* Within the overall sanitation system, there are 16 stream discharge units, two spray irrigation units, and one evapotranspiration unit. The remainder of the units rely on subsurface tile field discharge. In addition, four families are using recycled wastewater from a single installation.

The sanitary district employs a licensed sewage treatment plant operator to monitor, service and test all equipment in the project. Each unit in the project has a control panel which will alert the homeowner of any malfunction. Should a malfunction occur, the project operator is on call to handle emergencies.

---

*The Cromaglass model designated C-5 has not been listed by NSF and is not authorized to display the NSF seal.

INSTALLATION AND RESULTS

Two years ago, at the first NSF Onsite Conference, it was stated that "What we are trying to prove is that home aerobic systems seem to be a viable alternative to municipal treatment facilities in places where it costs $8,500 per house to put in a municipal treatment plant." The commission feels that an objective analysis of the equipment now in use in Boyd County must support the contention that this equipment is a workable alternative. The test results achieved thus far at Boyd County indicate a remarkable similarity to the tests conducted here at NSF. The conclusion drawn from this fact is that, despite the very different and fluctuating conditions encountered in the field, the system approach concept with regular inspection and maintenance can, and has, assured optimum operation of the various installations.

It was necessary in many cases in Boyd County to use **stream discharge** for the disposal of treated effluent, because, with the limited size or layout of the homesites involved, surface disposal was the only way these families could be served. The equipment which has been in service for many months now has consistently met or exceeded EPA **stream discharge** requirements. After initial treatment, all steam discharge units at Boyd County are followed by sand filtration and disinfection. It should be kept in mind that all homes which now are using surface discharge were previously either dumping raw sewage into Upper Chadwick Creek directly, or allowing septic tank runoff to flow into the creek.

For the surface discharge systems which have been in operation long enough to gather results (most of these are Multi-Flo), some rather consistent patterns have developed. The following table is a composite of surface discharge test results over the last five months:

DO, mg/l	pH	Temp, °F	SS, mg/l	BOD, mg/l
0.5-8.0	6.24-7.88	78-90	1-44	2-11
			80*	47*

*The results of one test following a unit malfunction.

Naturally, some units have performed better than others, and equipment malfunctions have occurred. The main equipment problem has been the failure of electric pumps. To date, the operator has had to replace nine malfunctioning pumps, all of which were under warranty. However, such malfunctions do show clearly on test results. Specific instances at Boyd County have yielded test results with suspended solids counts that range as high as 358 mg/l on subsurface discharge units.

Two other components of the project deserve special mention: evapotranspiration and wastewater recycling. At the O.T. Carter residence in Boyd County, with the assistance of the Cromaglass Corporation, the sanitary district has constructed a 2000-square foot evapotranspiration (ET) bed for the disposal of effluent from a Cromaglass model C-5 aeration plant. The ET system, which is actually two 1000-square foot beds, is sealed with plastic to prevent the high groundwater at the site from flooding the beds. Constructed with 8 inches of gravel and 18 inches of sand, the beds are crowned to facilitate rainwater runoff. Covered with a layer of topsoil, the beds have been planted with grass and junipers.

One of the values of an in-the-field test of such equipment is to observe the system's reaction to changing circumstances and shock loading. In the case of this particular evapotranspiration system, the design was intended to serve the needs of a family of four; however, because of a tragedy in the family, seven people now live at this site, including three small boys. While the result has been a large increase in water usage, particularly for laundry use, the evapotranspiration bed has thus far performed extremely well with only a slight modification to the distribution box. Prior to the installation of the treatment plant and ET bed, raw sewage stood in the yard of this house from an inoperative septic tank and drainfield, although the water usage was much lower than it is today. Although the high rainfall in the area caused some doubts as to whether the evapotranspiration concept would function properly, the results thus far have been very satisfactory. However, any final judgment on this installation should await monitoring of its performance through the winter and spring months ahead.

One of the most important and perhaps most controversial components of the Boyd County project from the outset has been wastewater recycling. This component appears to have been controversial to nearly everyone except the families involved in the demonstration project. In fact more requests were received for recycling equipment within the district than could be met. Because the use of this equipment was a source of considerable debate, it is particularly gratifying to find that it has been one of the most successful components of the project. Four recycle systems, serving five homes, are part of the overall Boyd County system; three Multi-Flo units, and one Cromaglass unit. At this time, however, test data are available only on the Multi-Flo equipment.

The recycling systems at Boyd County are composed of a treatment plant, holding tank, disinfectant, polishing filter and pressure tank. The standard treatment plant is followed by a 1000-gallon holding tank used to regulate flow by assuring an adequate quantity of relatively clean water for the recycling equipment, even in the event of a temporary problem with the treatment plant. From the holding tank, water is pumped past an iodine disinfectant, receiving a constant dose of 0.5 ppm. A small contact tank is used to retain the iodine-treated water for 20 minutes to allow for maximum disinfection. From here the water is filtered in a charcoal column equipped with automatic backwash to reduce maintenance. The charcoal removes iodine from the water, and provides a final polishing cycle by further reducing suspended solids before the water enters the pressure tank, ready for reuse.

Tests show that this recycle system provides extremely consistent results. A clear, odorless water of excellent quality is produced with suspended solids averaging 5 mg/l, and a zero fecal coliform count. Equally important has been the high degree of consumer satisfaction with the day-to-day use of recycled water.

Also at the first NSF Onsite Conference, it was stated that ". . . one of the most significant aspects of the system approach for the future is that if we form a sewer district, a body of municipal government, it gives us for the first time a vehicle by which the federal government can participate in the funding of home onsite sewage treatment plants." The goal was first to show that the sanitary district or system approach was a workable solution for the management of a rural system of onsite equipment, then to work with other federal agencies to provide grant funds for such systems. Although the Boyd County testing program will continue for some time, the initial results indicate that the system approach is indeed a workable idea for rural areas. Through the efforts of Senator Randolph and his excellent staff, and the many people at EPA who have expressed interest in the project, the ongoing funding for projects using the system approach is now a reality.

LESSON LEARNED

At this point, with the option for federal funding before us, it is important that we also point out some of the pitfalls and lessons of the Boyd County project. One of the most important and difficult problems in Boyd County has been simply getting the system built. Two years ago it was expected that, by this time, the demonstration would have been completed. Today, however, systems serving 24 families are now in place, with an additional 11 units now in the installation process. In the next month, an additional group of six installations will be made. There are three basic reasons for this slow pace of project completion, and each is important to the success of future systems. The first problem has basically been one of grants management, and has resulted in no small measure from unfamiliarity with the complexities to be encountered on the part of both the commission and the grantee. This has now been resolved through negotiations with the grantee. Basically, this is an internal matter that other federal agencies can avoid through careful preparation of grants management guidelines.

The second problem was one of legal delay. As with many new ideas, there has been a cautious attitude on the part of state and local regulatory officials who are responsible for public safety and health. Although frustrating at times, this cautious attitude, when viewed from a long-range perspective, is important to avoid serious mistakes involving the well-being and safety of the general public. Therefore, a great deal of time was spent in securing the necessary approvals to begin the implementation of the project while insuring proper safeguards. With more systems being constructed with federal grant funds, and the resulting familiarization with the advantages and limitations of alternative systems, this reluctance toward alternative systems should begin to diminish in the near future.

The third problem is that not many engineers and contractors are, as yet, familiar with standards, methods and requirements for the most efficient installation of alternative systems, particularly in the wide variety of installation problems found in servicing all homes in any given rural area.

Basically, there is a great need for the widespread availability of technical information on the design, installation and operation of alternative systems. The lack of an adequate technical base is one of the reasons that the Boyd County project has not been installed efficiently, resulting in duplications of effort and higher costs. It was found in Boyd County that it is a very complex task to install a system of individual units, and general expertise in this field is not yet available. There is a definite need for a detailed engineering study which takes into account the particular needs of each family being served, including family size, site layout, appliances in the home and many other factors.

It is important to note that wherever possible, *i.e.*, where pressure was available, iodine was used as a disinfectant rather than chlorine. Thus far in the program, we are satisfied that iodine is very reliable. Finally, it has been found that, where applicable, multifamily units offer greater economy of installation and operation, and greater efficiency for maintenance purposes.

OUTLOOK FOR THE FUTURE

Looking to the future, now that we can see the emergence of alternative systems as a recognized, accepted tool for addressing the problems of rural sanitation, we are hopeful that other federal agencies will follow the lead of EPA. This is particularly true of the Farmers Home Administration

(FmHA), which over the years has been so responsive to the needs of rural America. The Farmers Home Administration has a total of $800 million for water and sewer system construction ($200 million for grants and $600 million for loans). Unlike other major federal programs, FmHA is specifically charged with meeting the needs of rural America. Involvement in the system approach alternatives will help that agency serve more people with the resources that it has available.

It is essential at the outset that we proceed wisely so as not to abort the new grant process in its infancy.

For both the industry involved in alternative systems and for regulatory officials, it is extremely important that a comprehensive manual be developed (similar to the *Manual of Septic Tank Practice*) which will put into the hands of sanitarians, engineers and health officials an authoritative "How To" book on methods, standards and procedures. Such a manual, based on the experiences of the Boyd County project and the numerous other installations around the country, should provide specific information with respect to system applications, uses of disinfectants, surface and subsurface disposal system construction, recycling, etc. The value of such a manual cannot be overstated in the development of future systems, both as a guide and as a statement of minimum standards. In fact, the need for such a manual is so important to the whole concept of rural sanitation in Appalachia and elsewhere, that I would like to announce that the Executive Director of the Appalachian Regional Commission, Mr. Harry Teter, Jr., will within the next few months request the participation of industry, government funding agencies, and experts in the field to gather in Washington, D.C. to discuss the funding and composition of a representative committee to assist in the development of such a document.

Secondly, I urge the industry to establish a trade association which can effectively set industry standards and can present the industry's viewpoint in the drafting of future legislation and regulations which affect rural sanitation.

Further, with respect to future federal grants for a system approach, it is equally important that the parameters for funding alternative onsite equipment be structured in such a way as to insure that the end-product will resemble the intent, namely, to improve environmental health conditions in rural areas by providing sanitary services which are effective and within the financial reach of rural areas which have incomes below the national average. Obviously, no matter how good a proposed solution (or tool) may be, it is worthless if the intended users cannot afford it. It must be kept in mind during this discussion, that we are referring to relatively small systems serving perhaps less than 250 families. Therefore, the commission urges the consideration of the following funding proposals by other federal agencies:

First, the initial construction grant should provide not only for the purchase and installation of equipment, but should provide funds for an initial 90-day startup period for the system. This startup is a critical period of adjustment in which numerous unforeseen problems may arise requiring a great deal more maintenance than the normal operating period. Funded as a component of the basic construction grant, this 90-day period would assure adequate attention to the equipment without depleting the resources of the newly formed sanitary district with very limited capital. During this 90-day startup period, the sanitary district should be encouraged to collect its established maintenance fee so that, at the end of the startup period, the district will have sufficient operating capital to provide quality services on better than a marginal financial basis.

Another essential element of the initial construction grant should be the ability to stock spare parts and tools. This is essential from both the standpoint of starting the sanitary district off on the right foot as a financially self-sustaining, ongoing service organization, and from the standpoint of providing efficient maintenance. For example, the operator of the system will not have time to stand out in the rain trying to determine why a pump or other pieces of equipment are not working. Provided with an adequate parts inventory, he can simply replace nonfunctioning equipment and examine it for possible repair at a later date, as time permits.

Finally, and perhaps most controversial, as part of the initial grant for equipment and installation, the federal agency regulations should provide for *one* service vehicle for a community initiating a "system approach" concept. Such eligibility should be restricted as to the maximum dollar amount by a sliding scale based upon the number of families to be served, within an overall maximum and minimum size. This eligibility should be on a one-time only basis with an explicit prohibition against replacement.

The point of these recommendations on future federal funding is to insure that our ultimate purpose will be attainable. The purpose—affordable, effective sanitation for rural areas. Whatever we can do at the federal level to get new sanitary districts using a system approach with onsite equipment off to a good strong financial start, will serve us well for years to come. We are on the verge of a new era in meeting the needs of rural America. Let's start here to assure that the challenge ahead will be successfully met.

REFERENCE

1. "Senator Randolph Stresses the Need to Explore New Sewage Treatment Concepts," *Congressional Record* 122:152 (October 26, 1976).

26

ONSITE WASTEWATER FACILITIES FOR SMALL COMMUNITIES AND SUBDIVISIONS

Richard J. Otis

    Sanitary Engineer
    College of Engineering
    University of Wisconsin-Madison
    Division of Economic and Environmental Development
    University of Wisconsin-Extension
    Madison, Wisconsin 53706

INTRODUCTION

In 1970 approximately 19.5 million households or nearly 30% of all housing units in the United States disposed of their wastewaters by some form of private sewerage facilities.[1] This number is growing at an increasing rate, due to an emerging trend of population movement to rural areas where community sewage treatment facilities are not usually available. Retired persons are moving back to rural areas, as well as young families who are following the growth of industries on the outlying fringes of metropolitan centers.[2] Most of these rural households utilize septic tank systems to dispose of their wastewater. Because of poor design, construction or maintenance, however, a large number of these systems are failing to provide adequate treatment and disposal of their sewage.

Many households, while located in rural areas, are situated in small communities or subdivisions ranging in size from a few households to a hundred or more. In such instances, failing septic tank systems which allow raw or poorly treated sewage to reach the ground surface, surface body of water or even the groundwater create a severe public health hazard and nuisance because of the close proximity of homes. Public wastewater facilities are often the only solution to abate the problem.

CONVENTIONAL PUBLIC FACILITIES

The traditional method of providing public wastewater facilities is to construct a system of gravity collection sewers which convey all the wastewaters to a single community treatment plant. This "central" system is preferred by governmental authorities, engineers and the public alike for several reasons. First, the gravity sewer system is tried and proven. There is much technical expertise in the theory, design and operation of central sewerage which has led to great confidence in the system. Second, central sewerage is usually more cost-effective because of economies of scale. It is less costly to serve many people with one system rather than each one individually. Third, central sewerage allows ready application of central (and usually public) management which is responsible for the proper functioning of the system. The availability of a single entity to manage the system is quite desirable from a regulatory authority's viewpoint because the authorities have an entity against whom they can bring administrative or judicial action to abate water pollution problems. Central management is also favored by the homeowner who no longer has to worry about his private system.

For smaller communities and subdivisions, however, such a conventional collection and treatment facility is impractical because of the financial burden it places on the residents or developer. This is due largely to the high cost of collecting wastewater from each home or business. Smith and Eilers[3] computed the 1968 national average of total annual costs of municipal wastewater collection and treatment facilities which showed that 65% of the total annual cost is for amortization and maintenance of the collection system. A more recent study by Sloggett and Badger[4] of 16 small communities in Oklahoma showed a similar distribution (see Table I). It is clear from this breakdown of the total annual costs that the collection system is the most expensive component of any system.

Table I. Distribution of Total Annual Costs for Municipal Wastewater Collection and Treatment Facilities

	Amortization Cost		Current Expenses			
			Operation and Maintenance			
	Collection	Treatment	Collection	Treatment	Overhead	Total
Smith & Eilers (1968)[3]	60.3%	15.3%	4.7%	8.4%	11.3%	100.0%
Sloggett & Badger[4]	-- 72.6% --		14.2%	3.2% (lagoons)	10.0%	100.0%

In small communities homes are typically scattered, which causes the costs of sewering to rise dramatically. In their study of 16 wastewater collection and treatment systems in Oklahoma, Sloggett and Badger[4] showed that the costs per customer rise as the number and density of customers declines. Construction costs per customer were compared to the density and number of customers served (see Tables II and III).

Table II. Cost of Construction per Customer Relative to Density of Customers for 16 Community Wastewater Facilities in Oklahoma[4]

	Customers per Mile of Sewer			
	Under 30	30-39	40-49	Over 50
Number of systems	5	5	1	5
Average cost/customer (1972 dollars)	$1,100	$847	$696	$575
Average number of customers	96	119	310	256

Table III. Cost of Construction per Customer Relative to Number of Customers for 16 Community Wastewater Facilities in Oklahoma[4]

	Number of Customers Served			
	Under 100	100-199	200-299	300-400
Number of systems	6	4	3	3
Average cost/customer (1972 dollars)	$1,000	$798	$594	$434
Customers/mile of sewer	28.3	37.8	49.4	55.2

Both factors were shown to have a significant effect but the density of customers was shown to have the largest impact on per capita construction costs. In smaller communities where homes tend to be more scattered this cost can become excessive. Costs can reach $8000 per household for the capital portion alone and may be even higher if treatment beyond secondary is required to meet water quality standards. It is not unusual for the cost of the complete system to approach the total equalized value of the community.[5]

To help communities meet the water quality goals of the Federal Water Pollution Control Act Amendments of 1972, the federal government was authorized by a provision in the Act to provide grants in aid of construction for 75% of the grant-eligible portions of the wastewater facility. The availability of these grants would help offset the high per capita costs in small communities but, unfortunately, small communities have difficulty in obtaining them.

The federal funds are allocated to the individual states on the basis of need, but each state is given the power to determine how the funds are to be spent. Only minimum requirements are set out by the Act for states to follow in preparing a priority list of projects. For example, the Act requires that consideration be given to the severity of the pollution problem, the population affected, the need for preservation of high quality waters and national priorities. The federal regulations seem to give the states some discretion by not requiring strict adherence to their rankings of pollution discharges. Thus, the priority lists usually work to the disadvantage of small communities, in that many of them are near the bottom, preceded by communities with larger populations and larger pollution discharges. This emphasis denies small communities any expectation of receiving badly needed funding for public facilities in the near future. It is obvious from this discussion that it is impractical to expect many small communities to construct conventional public wastewater facilities to eliminate failing private systems.

Central sewerage is also very costly for subdivisions. Because of the large front-end cost of installing conventional gravity sewers with no immediate return, developers prefer to utilize private septic tank systems to dispose of the wastewater. In older subdivisions, septic tank systems were often installed in unsuitable soils or constructed improperly, resulting in mass failures. The only solution has been to extend interceptor sewers to the subdivision from the nearby municipality.

This becomes a very costly proposition in several ways. The lack of an alternative forces the developer to subdivide land with relatively good soils. Not only does this often remove good agricultural land from production, but it also results in scattered development about a metropolitan center. The development will increase the tax base of the local government, but the scattered development also increases the costs of providing other community services, such as roads, police and fire protection and other utilities. If septic tank system failure requires eventual extension of municipal sewerage for the outlying subdivisions it becomes extremely costly, and may also be undesirable in some cases because of the strip growth that often occurs along the interceptor routes. The result may be a net economic and environmental loss to the community.

The need for a cost-effective yet viable alternative is certainly indicated. Regulatory officials and engineers are realizing that if the goals of the

Federal Water Pollution Control Act are to be met, more practical facilities must be developed for small communities and subdivisions.

NONCENTRAL FACILITIES AS AN ALTERNATIVE

A "noncentral" facility of several treatment and disposal systems serving isolated individual residences or clusters of residences may offer a less costly alternative to the conventional central facility in the nonurban setting. As Table I indicates, approximately two-thirds of the total annual cost of a conventional facility is attributable to the collection system. In a community of scattered homes this proportionate cost could be even higher. If the central treatment plant could be eliminated, long sewer extensions collecting wastes from widely spaced homes would not be necessary. Instead, treatment and disposal could be provided where the wastes are generated. Individual or jointly used septic tank systems or other treatment and disposal methods could be used. Such a noncentral facility of disperse systems could result in a substantial savings.

The implementation of a noncentral facility would not exclude the use of central management, which is an extremely attractive factor of conventional community facilities. Though a relatively untried concept, central management of a noncentral facility could be employed. In fact, central management would be crucial to its proper functioning.

The noncentral facility offers several advantages over the central sewerage approach:

1. Existing functional septic tank-soil absorption systems can be utilized rather than providing new service. Often, homeowners who are not having trouble or who have recently installed new septic tank systems do not wish to support community action that will cost them more money unnecessarily. Incorporating existing systems into the public system minimizes this opposition, as well as reducing the total cost of the public facility.

2. Isolated single homes and clusters of homes can be served individually instead of extending costly sewer lines out to them. This could be equally advantageous to existing communities, as well as newly platted subdivisions. Where future growth is not expected to be great enough to warrant sewer extensions, individual septic tank systems could be used. In cases where substantial growth is expected, such as in newly platted subdivisions, the first few homes built could be served by holding tanks which would be pumped and maintained by the management entity. When the number of homes increased to the point where a common disposal system is warranted, it could be built on land reserved for that

purpose. This would delay construction until the time there are enough contributors available to pay for it.

3. Less costly treatment facilities can usually be constructed. In addition, subsurface disposal can often be employed which requires minimal treatment and avoids the necessity of upgrading the treatment plant to meet changing standards for effluent discharges to surface waters. Where subsurface disposal is not possible the smaller flows may allow other simple treatment methods to be used. In addition, by limiting the area served the necessary excess capacity required for future growth is accurately known, providing a more optimal design.

4. A more cost-effective facility may encourage smaller communities to proceed with construction rather than waiting for federal construction grants. This would speed abatement of water pollution problems. Where financial aids are necessary, a greater number of community facilities could receive construction grants because of the fewer dollars required for each project.

5. More rational planning of community growth is possible. Strip growth, which is encouraged by the construction of interceptor sewers used to collect wastes from outlying clusters of homes, could be avoided. Growth could be encouraged in the more desirable areas by providing public service in those areas only.

6. Noncentral facilities are more ecologically sound because the disperse systems dispose of the wastes over wider areas. Through this practice the environment is able to assimilate the waste discharge more readily, which reduces the need for mechanical treatment and the associated energy consumption.

Of course, there are disadvantages to noncentral facilities which must be overcome if this alternative is to be successful:

1. Central management of a facility of small disperse systems is a fairly new and untried concept. Methods of public ownership of systems on private land, necessary for proper operation and maintenance, must be tested. Operation and maintenance costs also may be higher than for conventional facilities because of the "noncentral" nature. Because of the lack of experience, other problems will arise which may not be anticipated.

2. There is little public confidence in wastewater facilities that do not convey the wastewater from areas of habitation; therefore, a noncentral facility may be unacceptable. Failure of a conventional treatment plant is easy to accept, because it is usually a safe distance from any homes and does not disrupt the household routine. However, if failure of a

treatment and disposal system within a noncentral facility occurs, repairs must be made immediately.

3. Provision for the community's future growth is more difficult. A small reserve capacity can be built into each system which serves an area with some undeveloped lots, but if a landowner wishes to build where public service is not yet available, a decision must be made as to whether service should be provided. Because providing public service to single homes one at a time can be costly, a choice must be made between constructing individual systems, providing holding tank service until more homes are built in the area, or immediately constructing a larger joint system to handle anticipated growth.

4. By present guidelines many components of a noncentral facility may not be eligible for grants in aid of construction, which are available through various federal and state financial aid programs. This would have the effect of increasing the cost to each customer served in comparison to central sewerage, even though the total costs may be less.

In general, the potential of noncentral facilities seems to warrant further investigation. Many of the possible shortcomings of this alternate facility may vanish as some are constructed and experience gained.

COLLECTION AND TREATMENT ALTERNATIVES FOR NONCENTRAL SYSTEMS

Proper facilities planning involves a systematic comparison of alternative methods of dealing with a wastewater treatment and disposal problem. The purpose of this comparison is to identify the most "cost-effective" solution which will minimize total costs to society over time. These costs include monetary and environmental, as well as other nonmonetary costs.

The commitment by regulatory agencies and engineers to conventional gravity sewers with a common central treatment plant, however, has eliminated many worthy alternatives from consideration. If this bias can be changed, the utilization of the noncentral concept has the potential of significantly reducing the environmental and monetary costs of wastewater facilities in many communities by either reducing the size or eliminating the collection system altogether and by simplifying the treatment facility.

The most extreme noncentral system would be one where each home and other establishment were served by its own individual septic tank system. However, the most cost-effective community system will usually lie somewhere between the two extremes of central sewerage and individual systems. Either because of economies of scale or because site conditions are unfavorable for individual disposal systems, joint systems serving several homes may be constructed. The end result may be a mix of several individual and joint systems.

Collection Alternatives

The single most expensive portion of central sewerage is the gravity collection system, yet alternatives to it are rarely evaluated. Three interesting alternatives might be employed.

Small-Diameter Gravity Sewers

To take advantage of economies of scale or to avoid adverse sites, a nearby area might be available for construction of a joint system. In such cases, gravity sewers can be used to collect and convey the wastes to the disposal site. To reduce the costs of conventional gravity sewers, small-diameter (4-inch) pipe offers an alternative if septic tank effluents rather than raw wastes are collected. The collection mains are joined by a typical gravity house connection coming from a septic tank or, in those instances where the elevation of a property would make it difficult to be served by a gravity system without a large cut, the building would be provided with a pump located in a chamber immediately following the septic tank to elevate the effluent into the system.

The individual septic tanks would provide partial treatment of the wastewater by removing the larger solids to allow the use of smaller-diameter pipe for collection. Because sand and other grit also would be removed in the septic tank, normal cleansing velocities in the mains need not be maintained. The 4-inch-diameter mains are installed at a minimum gradient of 0.67%, based on a minimum velocity of 1.5 feet per second at half pipe capacity.[6] Under these conditions a 4-inch-diameter pipe can carry over 2000 gph sufficient to serve 670 persons, assuming a peak flow of 3 gph per person.[7] Regular flushing to provide cleansing velocities (greater than 2.5 feet per second) can be provided if necessary by collecting and pumping the effluent from several homes at the upstream end of each main for periodic surcharging. This type of collection system has been used extensively in South Australia since 1962 without surcharging with very good results.[6]

Pressure Sewers

Where topography or soil conditions make gravity sewers costly, pressure sewers may be more economical. Pressure sewers have been tried in several places and have performed favorably.[8-10] This system consists of a septic tank at each building or cluster of buildings to remove the large solids from the wastewater followed by a pump chamber with a small submersible pump to pump the settled effluent from the septic tank into a small-diameter plastic main. (A grinder pump can be used to grind and pump the raw waste rather than using a septic tank.) Construction costs are reduced because the pipe need not be laid at a specific grade but can follow the contour of the land with the main being located just below the frost line. This permits the use of a simple trenching machine and eliminates deep cuts, often necessary for gravity sewers.

Flexibility for growth is good because the smallest pipe size used can accommodate up to 20 homes before a larger main is necessary. Further pipe size increases are necessary only by 30 dwelling increments.[8] Such a system can easily be designed to handle nominal growth.

Vacuum Sewers

In temporate regions where soils are shallow, vacuum sewers offer another alternative. They provide many of the advantages of pressure sewers.

Treatment and Disposal Alternatives

The degree and method of wastewater treatment depends upon the constraints of the receiving environment. In large municipal systems the volume of wastewater collected usually requires that a stream or river be used as the receiving environment. To protect the water quality, secondary or higher degrees of treatment are necessary. In addition, mechanically intensive treatment methods must be employed because large land areas are often unavailable. Thus, treatment and disposal alternatives for larger municipalities are limited and costly.

In small communities using noncentral facilities, however, simpler and less costly treatment and disposal methods can be employed. Smaller volumes of wastewater permit a wider choice of disposal methods which may require only minimal treatment. Rather than discharging to a surface body of water, land disposal or evapotranspiration may be a more cost-effective alternative.

Land Disposal Alternatives

Land disposal becomes a particularly attractive alternative in rural areas where land is more likely to be available. The soil is an effective treatment and disposal medium which should be utilized whenever possible. One type of system readily adaptable to noncentral facilities is the septic tank-soil absorption field. This system can be designed to dispose of wastes from single homes or large clusters of homes. Unfortunately, these systems are usually considered only as an interim method of treatment and disposal until sewers are available.[11] This is because septic tank systems have not been understood and, therefore, have been improperly designed, installed and maintained. Thus, many failures have resulted which have created a lack of confidence in their reliability.

Recently, however, practical design criteria and installation procedures have been developed for soil absorption systems.[12,13] If followed, septic tank systems can be expected to last 20 years or more under many soil conditions. If used under the rubric of central management, these systems should be seriously considered as a realistic alternative.

Other systems, such as spray or ridge and furrow irrigation and infiltration-percolation ponds, also utilize the soil for treatment and disposal. These may be viable alternatives to consider depending on climate, yearly distribution of wastewater flow, etc.

Surface Water Discharge

Where soils are unsuitable or sufficient land is unavailable, it is necessary to use other methods of disposal. As an alternative, surface waters may be used as the receiving environment. This often requires that higher levels of treatment be provided prior to wastewater discharge. Intermittent sand filters, a method of treatment which was abandoned because of its requirement for large land areas, offer an alternative to package plants or lagoons. In noncentral systems sand filtration becomes viable because of its simplicity and reliability. The smaller wastewater flows to be treated and higher loading rates reduce the required land area. High-quality effluents low in biochemical oxygen demand and suspended solids can thus be produced with a minimum of maintenance.[14,15]

Evapotranspiration

In climates where evapotranspiration exceeds precipitation, evapotranspiration provides another disposal alternative. If land is available but is unsuitable for soil absorption, this alternative, which has low treatment requirements, may be more cost-effective than discharging to surface bodies of water where high degrees of treatment are necessary.

Other

To reduce the size and cost of any treatment and disposal facility, chosen waste segregation and water conservation may be built into the system. Water conservation can reduce the total volume of waste to be disposed of while separate handling of toilet and greywater wastes may simplify treatment.

It may be that after consideration of these and other alternatives central sewerage is the best solution. Generally speaking, however, most small communities can make use of one or a combination of alternative systems that may be a mix of individual and joint systems all under public ownership to provide the most cost-effective facility. Public ownership would make many components of each system within the facility eligible for construction grants and provide proper and reliable maintenance needed for long-life systems.

WESTBORO, WISCONSIN CASE STUDY

Few viable alternatives for small communities and subdivisions have ever been tried nor have there been incentives to do so. However, with the emphasis for cleaning up our nation's waterways moving from large municipalities to small communities, it is being realized that conventional solutions are too costly and may not be practical. This has caused the search for alternatives by several communities including Westboro, Wisconsin,[16,17] Glide, Oregon[18] and Fountain Run, Kentucky.[19] The Glide, Oregon and Fountain Run, Kentucky plans are discussed elsewhere at this conference.[20,21]

Description of Westboro

Westboro, Wisconsin is typical of hundreds of small rural communities in the Midwest that are in need of improved wastewater treatment and disposal facilities but are unable to afford conventional sewerage. Westboro was established as a permanent northern Wisconsin community in the late 1850s as a result of the lumber industry (Figures 1 and 2). By 1900 the population had grown to about 900, but with the decline of the lumber industry the population also declined. The present population is approximately 200 persons. A small machine tool company and a sawmill employing a total of 5 to 10 people remain in town.

The community of Westboro has no municipal wastewater collection or treatment facility. There are 94 buildings located in the community, of which 69 are occupied, including a school, four churches and several commercial establishments. All are served by private wastewater disposal systems. A 1971 survey by the Wisconsin Department of Natural Resources (DNR) showed that 80% of the septic tank systems were discharging wastes above ground. Many of the systems were found to be interconnected by common drains discharging directly into Silver Creek which flows through

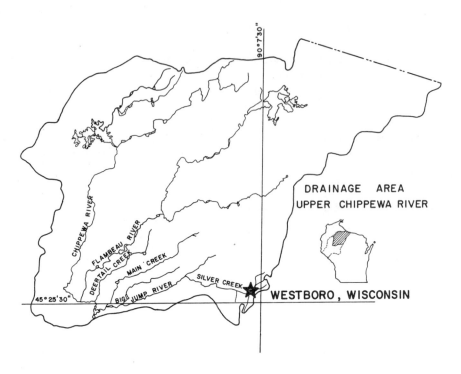

Figure 1. Geographic location of Westboro, Wisconsin.[16]

town. This situation was declared a nuisance and a menace to health and comfort, as well as the public rights in the Upper Chippewa River Basin. Consequently, DNR issued an order to Westboro to stop all private homes from discharging wastes into Silver Creek, either by upgrading all failing septic tank systems or constructing public wastewater facilities.

Proposed Central Sewerage

The soils and lot sizes prevent the replacement of most of the failing septic tank systems on an individual basis (Figure 3) so a public facility was determined to be necessary. The community formed a sanitary district, "Sanitary District No. 1 of the Town of Westboro" (Figure 2) and in 1967 contracted with an engineering firm to complete a facilities plan to abate the water pollution problem. The firm investigated two alternatives: (1) gravity collection to an extended aeration package treatment plant, and (2) gravity collection to a two-cell lagoon. Both plans served only 60 of the 69 occupied buildings. Homes to the north of town near

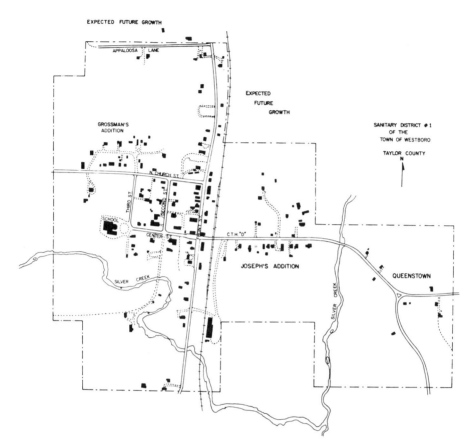

Figure 2. Sanitary District No. 1 of the Town of Westboro.[16]

Appaloosa Lane and those east of Silver Creek in Queenstown were not included (see Figures 4 and 5). Construction costs updated in 1976 were $135,700 for the collection system required for both alternatives, $115,650 for the package plant with a required 30-day effluent holding pond and $117,925 for the stabilization lagoon.[16] Total construction costs of these facilities, therefore, are estimated to be $251,350 for Alternate 1 and $311,625 for Alternate 2, including engineering and contingencies.[16]

The Westboro Sanitary District applied for federal EPA grants in aid of construction, but their priority for receiving funding is very low. As of February 1976, Westboro was 318 on the list of 420 to receive 75% of eligible costs of construction of the treatment plant and interceptors and 398 to receive similar funding for the sewers. This virtually rules out the possibility of obtaining a community facility for several years.[22]

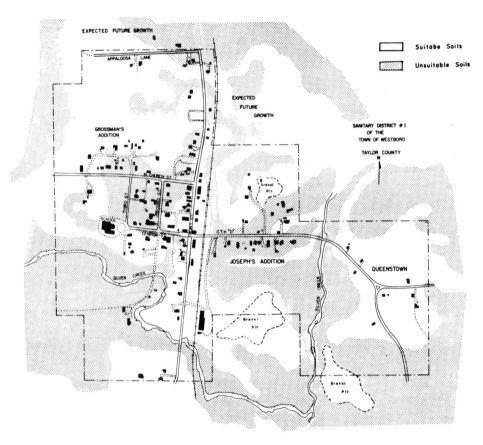

Figure 3. Soil suitability for conventional septic tank systems.[16]

Alternative Noncentral Facilities Evaluated

Having a sincere interest in abating their problem the residents of the Westboro Sanitary District agreed to cooperate with the Small-Scale Waste Management Project at the University of Wisconsin to develop an alternate plan which might be a more cost-effective facility. The objectives of the project were to evaluate the use of several small treatment and disposal systems placed in strategic locations within the community to serve individual homes or clusters of homes, but under central management, to compare total costs of alternate plans to the proposed conventional facility, and to determine the best method for management of such alternate facilities.

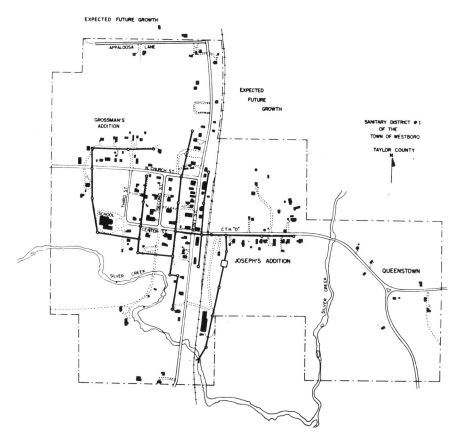

Figure 4. Central System Alternate 1: conventional gravity sewers—extended aeration treatment.[16]

Because the collection sewers represented approximately half of the total construction costs in the conventional plans, an effort was first made to reduce the size of the collection system. The community was divided into natural groupings of buildings for the consideration of various alternatives. Five groupings were made: (1) Front Street area, extending from Silver Creek north to the cemetery and from Second Street to the railroad tracks, (2) Grossman's Addition, including the area west of Second Street and the school, (3) Joseph's Addition, (4) Queenstown and (5) Appaloosa Lane, including the scattered houses north of the Front Street area (Figure 6). Each area was considered separately and in combination with adjacent areas to develop the most cost-effective system.

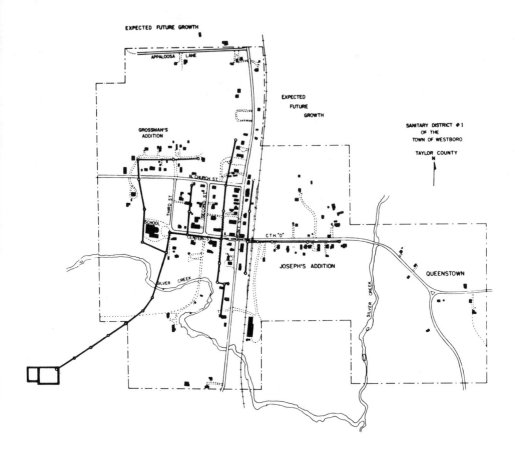

Figure 5. Central System Alternate 2: conventional gravity sewers—lagoon treatment.[16]

Collection systems were considered to be the best alternatives for the Front Street area which includes the business district. This area is primarily divided into small 150 ft x 50 ft lots. Most of the lots are developed leaving little area to construct new individual septic tank systems. Joseph's Addition is a low-lying area with poorly drained soils. Individual mound systems could be installed, but a common system would be more cost-effective. A similar condition occurs in Grossman's Addition area where individual systems could be installed, but because of the density of homes, a common system offers the greatest advantage.

Several alternatives were considered for these areas. Because of the limited disposal sites available, it was appropriate to combine the Front

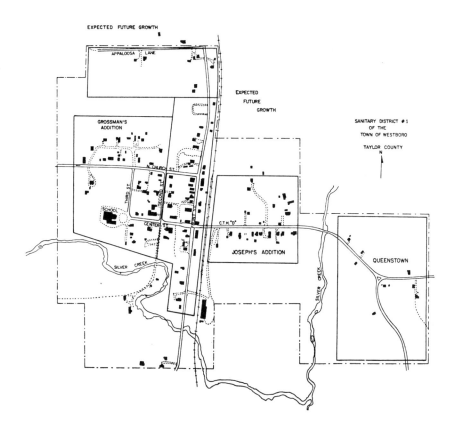

Figure 6. Grouping of homes made for evaluation of alternatives.[16]

Street and Joseph's Addition areas, with disposal to an extensive sand bench along Silver Creek east of town. Both pressure and small-diameter gravity sewers collecting septic tank effluents were evaluated for these combined areas. In Grossman's Addition, four alternatives were evaluated. Because of topography, collection by small-diameter gravity sewers to a point southwest of the school is well suited for this area. Disposal alternatives considered were soil absorption, sand filtration with chlorination before discharge to Silver Creek and pumping to the Front Street and Joseph's Addition gravity system. The fourth alternative was a pressure collection system, also combined with the Front Street and Joseph's Addition pressure system.

The remaining Appaloosa Lane and Queenstown areas are too sparsely developed to warrant collection systems. At present, individual systems seem to be the best alternative. Farm land with soils suitable for either a conventional or mound disposal system exist.

In summary, the noncentral alternatives evaluated were:[16]

Alternate 1

Part A: Grossman's Addition—Small-diameter gravity sewers discharging to a soil absorption field west of the school (design load of 10,000 gpd).

Part B: Front Street and Joseph's Addition—Small-diameter gravity sewers discharging to a soil absorption field northeast of Joseph's Addition (design load of 20,000 gpd).

Alternate 2

Part A: Grossman's Addition—Small-diameter gravity sewers discharging to a soil absorption field west of the school (design load of 10,000 gpd).

Part B: Front Street and Joseph's Addition—Pressure sewer discharging to a soil absorption field northeast of Joseph's Addition (design load of 20,000 gpd).

Alternate 3 (Figure 7)

Part A: Grossman's Addition—Small-diameter gravity sewers discharging to intermittent sand filters west of the school with chlorine disinfection before disposal into Silver Creek downstream from the community (design load of 10,000 gpd).

Part B: Front Street and Joseph's Addition—Small-diameter gravity sewers discharging to a soil absorption field northeast of Joseph's Addition (design load of 20,000 gpd).

Alternate 4 (Figure 8)

Part A: Grossman's Addition—Small-diameter gravity sewers discharging into intermittent sand filters west of the school with chlorine disinfection before disposal into Silver Creek downstream from the community (design load 10,000 gpd).

Part B: Front Street and Joseph's Addition—Pressure sewers discharging to a soil absorption field northeast of Joseph's Addition (design load of 20,000 gpd).

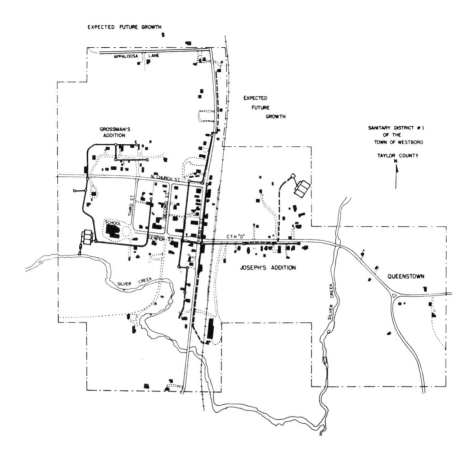

Figure 7. Noncentral Alternate 3.[16]

Alternate 5 (Figure 9)

Small-diameter gravity sewers serving all areas discharging to a soil absorption field northeast of Joseph's Addition (design load of 30,000 gpd).

Alternate 6 (Figure 10)

Pressure sewers serving all areas discharging to a soil absorption field northeast of Joseph's Addition (design load of 30,000 gpd).

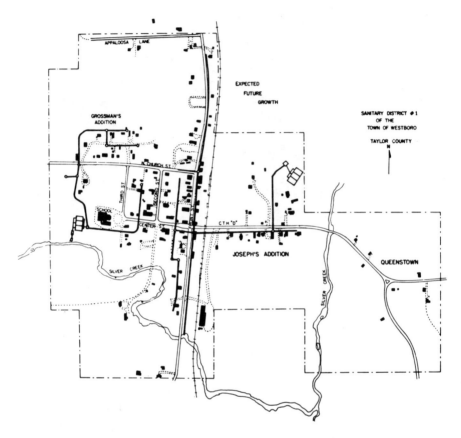

Figure 8. Noncentral Alternate 4.[16]

Facility Selection

Final selection of one alternative over several others depended on three criteria: environmental impact, total cost and system reliability. The first two are obvious, since it is the goal of the engineer to design a facility which will protect the environment for the least cost. Judgments must be made as to whether additional environmental protection warrants added facility costs, but much of this can be decided objectively. System reliability is less objective, however, and is influenced by the engineer's past experience. It is often more a confidence factor, which will eliminate some alternatives from consideration because they are not felt to be viable.

INDIVIDUAL ONSITE WASTEWATER SYSTEMS 265

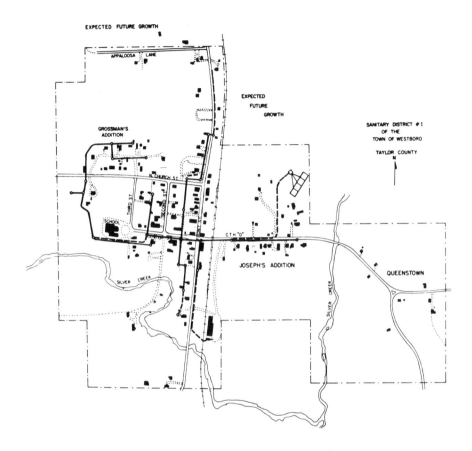

Figure 9. Noncentral Alternate 5.[16]

This factor is what usually eliminates septic tank systems from consideration. Each of these must be weighed in the final selection.

The "Noncentral" Alternate 5 was selected as the best facility after weighing each criterion, though some assumptions made in the analysis must be proven through experience. This facility is a system of small-diameter gravity sewers with final effluent disposal in a single soil absorption field (Figure 9). Pretreatment would be provided by individual septic tanks at each home. The effluent is conveyed to a conventional soil absorption field which is divided into 3 beds providing 1.5 times the estimated area necessary for absorption. Two beds would be in use at all times. The third would be alternated into use on a regular basis. This

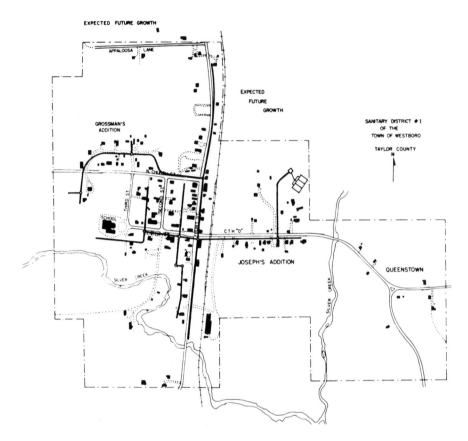

Figure 10. Noncentral Alternate 6.[16]

arrangement permits a bed to rejuvenate by "resting" and provides a stand-by unit. Homes outside the collection system would be served by individual septic tank systems.

This facility appears to be the least costly and more environmentally sound than the other alternatives evaluated. The reliability of this type of facility has not been established, however, but its selection is warranted because it is designed from extensive experience with smaller systems. In addition, its cost and environmental impact are a significant improvement over the conventional central facilities.

Cost comparisons between all alternates were made using present worth analysis. Present worth is equal to the initial cost plus the amount of money which must be invested at the present time to cover the costs of operation and maintenance over the life of the system. A lifetime of 20 years with an annual interest rate of 7% was used in these computations. A summary of the estimated present worth of each alternate is presented in Table IV.

Table IV. Summary of Present Worth Costs of Alternate Facilities[16]

"Central" System Alternate 1
  Extended Aeration Treatment Plant

Collection	$136,295.00	
Treatment	170,065.17	
Hookup	31,050.00	
Individual Systems	11,976.23	
		$349,386.40

"Central" System Alternate 2
  Raw Sewage Stabilization Pond

Collection	$136,295.00	
Treatment	185,528.00	
Hookup	31,050.00	
Individual Systems	11,976.23	
		$384,849.23

"Noncentral" System Alternate 1

Part A: Grossman's Add.–S.D. Gravity Sewers to Soil Absorption
Part B: Front St. & Joseph's Add.–S.D. Gravity Sewers to Soil Absorption

Part A	$124,454.64	
Part B	145,229.00	
Individual Systems	11,976.23	
		$281,659.87

"Noncentral" System Alternate 2

Part A: Grossman's Add.–S.D. Gravity Sewers to Soil Absorption
Part B: Front St. & Joseph's Add.–Press. Sewers to Soil Absorption

Part A	$124,454.64	
Part B	185,308.00	
Individual Systems	11,976.23	
		$321,738.87

"Noncentral" System Alternate 3

Part A: Grossman's Add.–S.D. Gravity Sewers to Sand Filters
Part B: Front St. & Joseph's Add.–S.D. Gravity Sewers to Soil Absorption

Part A	$148,038.00	
Part B	145,229.00	
Individual Systems	11,976.23	
		$305,243.23

Table IV, Continued

"Noncentral" System Alternate 4

Part A: Grossman's Add.–S.D. Gravity Sewers to Sand Filters
Part B: Front St. & Joseph's Add.–Press. Sewers to Soil Absorption

Part A	$148,038.00	
Part B	185,308.00	
Individual Systems	11,976.23	
		$345,322.23

"Noncentral" System Alternate 5

Total Gravity Sewers to Soil Absorption

Joint System	$254,440.00	
Individual Systems	11,976.23	
		$266,416.23

"Noncentral" System Alternate 6

Total Pressure Sewers to Soil Absorption

Joint System	$294,154.00	
Individual Systems	11,976.23	
		$306,130.23

To make a fair comparison of costs, the conventional central facility alternates were redesigned to conform with present regulations and site conditions. Private individual system construction estimates were also included for those homes not served by the conventional alternates. While the cost of replacing these septic tank systems would not be borne by the District in the case of the conventional system, their inclusion provides a fairer comparison between the "central" and "noncentral" alternates. Hookup costs are also included, for Alternates 1 and 2. They are estimated to be $450 per service connection. Hookup costs for the "noncentral" alternates are included in the construction costs.

"Noncentral" Alternate 5 is estimated to be the least costly of all the alternatives evaluated. The present worth of Alternate 5 is $266,416 or approximately $3861 per household, as compared to $349,386 or $5063 per household and $384,849 or $5578 per household for the "Central" Alternates 1 and 2, respectively. Thus, the noncentral system results in a 25 to 30% savings per connection over the conventional facilities.

The environmental impact of "Noncentral" Alternate 5 should be minimal. Only nitrogen in the form of nitrate is expected to leach through the soil to the groundwater in amounts that may be significant.

With the field's location near Silver Creek much of the nitrate will probably flow into Silver Creek increasing its nitrogen content. Phosphorus, however, will have been removed through adsorption and precipitation reactions in the soil.[23] Pathogenic bacteria and viruses should also be removed.[24,25] This method of disposal is superior to direct discharge of treated effluent into Silver Creek because such effluents contain phosphorus and pathogenic organisms and viruses, as well as nitrogen.

Institutional Arrangements

To properly manage its noncentral system, the Westboro Town Sanitary District must regulate all individual and jointly used onsite disposal systems operating within its boundaries. While no town sanitary district has attempted this in Wisconsin, it is within their power to do so.[16,17] Briefly, advantages would arise because the district would be able to better perform the following functions:

1. Design and construct sanitary facilities for existing and future structures.
2. Identify and obtain rights to land with suitable soils for disposal areas setting aside sufficient areas for future growth.
3. Operate and maintain all individual and joint systems within the district, including pumping of all septic tanks.
4. Monitor groundwater and surface water quality to detect failing systems.
5. Repair or reconstruct any failing systems.
6. Establish a fair assessment and rate structure for subscribers to pay for cost of services.
7. Apply for grants in aid of construction for portions of the sanitary facilities that the district will own.

Access to Private Property

Many of the facility components of the recommended noncentral facility, such as septic tanks and effluent pumps, will be located on private property. Since regular maintenance of these components is necessary for proper functioning of the facility, permanent legal access to the properties must be obtained for purposes of installation, operation and maintenance. These easements are required prior to construction. In most cases, however, the exact location of the existing septic tank is unknown. Therefore, a general easement tied to the location of the septic tank rather than the property line is proposed.[16] Easements must also be obtained for any collection sewers of joint systems which cross private property.

It is hoped that the necessary easements can be acquired voluntarily from the property owners. Since all property owners within the district will be assessed for the cost of the facility, whether they use the facility or not, the owners might be encouraged to grant the required easements. Another factor which might serve to encourage the property owners to grant easements is the risk of prosecution by the county or state against the continuing use of their failing septic tank system. If the property owner fails to grant the easements voluntarily, however, the district could condemn such easements through eminent domain proceedings. This alternative, of course, is undesirable. The success of the noncentral system depends on a strong "community effort."

Subscriber's Responsibilities

The district will be responsible for the operation and maintenance of all components of the facility located on private land commencing from the inlet of the septic tank. The property owner's only responsibility will be to provide and maintain the lateral drain from his home or establishment to the septic tank and any power costs associated with lifting his effluent into the collection sewer or absorption field, if necessary.

Financing of Proposed Plan

Since Westboro's priority for federal EPA construction grants is very low, other sources of funding were sought for construction of the proposed facility. Tentative commitments were obtained from the Wisconsin Department of Natural Resources and the USDA Farmer's Home Administration (FmHA) for grants totaling approximately 50% of the construction costs. The remainder of the construction funds would be provided by a FmHA 4%, 40-year loan.

Special easements and monthly charges will have to be determined by the commissioners of the Sanitary District. However, to estimate their grant contribution, FmHA assumed a monthly charge of $8 per residence, $15 per commercial establishment and $1240 for the school and a 0.004 sanitary levy which would be sufficient to retire the debt and cover costs for operation and maintenance. Special assessments of $200 per residence, $300 per commercial establishment and $1500 to the school would be the remaining contribution made by the community.

Because those residents who recently constructed new septic tank systems would be reluctant to join the system, credit would be extended to them depending on the age and condition of their septic tanks. In most cases the septic tank would be suitable for use by the community system, thereby saving the district the cost of a septic tank. This savings will be returned to the owner in an inverse proportion to the age of the tank.

New subscribers joining the system after construction of the facility should be expected to pay a larger assessment. A formula might be worked out whereby new residents would pay all costs of hooking to the collection sewer and their share of the absorption field. This is a decision which will have to be made by the district commissioners.

While the costs are within the financial capabilities of the community, the financial grants are not as large as hoped. Biases in funding guidelines prevent agencies from providing more despite the fact that Westboro made efforts to construct a more cost-effective facility. The DNR grant from funds provided by the state of Wisconsin is limited to 25% of construction costs of grant-eligible items. Any portion of the system located on private property, whether or not permanent easements have been given, is not considered eligible. This is unfortunate, because it disallows the septic tanks which provide partial treatment necessary to permit the use of less costly sewers. The savings made by DNR as a result of the more cost-effective facility are not passed on to the community. Land purchase is also excluded, though the soil becomes the final treatment facility in this plan.

The Farmers Home Administration does not distinguish between items for eligibility but rather bases its grant contribution on what it feels is the community's ability to pay. For the portion to be paid by the community, a 5%, 40-year loan is offered. The amount of the grant portion is determined by assuming a monthly charge and special assessment per residence and a sanitary tax levy according to the wealth of the community. This income is used to retire the debt and pay for operation and maintenance over the 40-year loan period. By back calculating, the amount of the grant is determined, but it cannot exceed 50% of the total construction costs.

Both of these policies do not provide much incentive for communities to construct more cost-effective facilities. The guidelines for the DNR grant program should be reevaluated to see whether or not vital portions of the system located on private property can be grant-eligible if permanent easements are obtained. If not, the community would be inclined to construct as much of the system on public right-of-way as possible. This could increase the cost to DNR and the taxpayer, but reduce the cost to the resident.

The FmHA policy provides little more incentive to construct less costly systems. By back calculating from a basic monthly charge and special assessment, the cost to the community residents changes little, regardless of the cost of the facility. This policy must be made more flexible to credit communities willing to make an effort to reduce costs.

Monitoring Program

Performance reliability of the proposed facility remains to be proved. Public ownership and management of septic tanks located on each private lot served is rather new. The success of small-diameter gravity sewers depends upon proper maintenance of the septic tanks. Further, the effects of a large soil absorption field on groundwater quality have not been established. These items will be monitored by SSWMP for the next three years, pending the availability of funding.

SUMMARY

The demand for less costly wastewater facilities for small communities or fringe areas is increasing. Regulatory officials and engineers are realizing that if the goals of the Federal Water Pollution Control Act are to be met, more practical facilities must be developed for small communities and subdivisions. Recent studies have shown that up to 25 to 50% savings can be realized in public wastewater facilities in small communities by using alternatives to conventional sewerage.[16,18,19]

Though the results of these studies indicate that significant savings can be made by investigating other alternatives to conventional sewerage, there are several deterrents to their widespread acceptance. Biases of engineers, regulatory agencies and funding agencies favor central gravity sewers and treatment plants. Probably one of the greatest deterrents to the use of such facilities is technical knowledge and experience with the performance of relatively untried techniques. Innovative designs take more time to prepare and have more risk associated with them. Because engineering fees are usually based upon a percentage of the construction costs, there is little incentive to be innovative. The engineer gets paid less for doing more work and at a greater risk. Facilities like these need to be constructed and monitored to gain familiarity with noncentral systems to increase their acceptance.

Regulatory agencies also favor conventional systems, because of confidence and familiarity in tried and proved methods. Innovative designs, therefore, take more time to review. Thus, the engineer is more likely to design a conventional facility that creates fewer stumbling blocks with the reviewing agency.

Another deterrent to acceptance of such facilities is the question of whether this type of plan would be eligible for federal and local construction grants. Certainly there is bias in favor of conventional sewerage, because of present component eligibility guidelines. Thus, while a conventional facility may be more costly because of its eligibility for construction grants, it becomes less costly to the subscribers. This bias is wasteful of tax dollars, as well as environmentally unsound, for it encourages communities to delay abatement efforts until funding is available.

Obviously, what is needed are additional planning studies of this nature, working with several communities or subdivisions each having different characteristics. Such studies would provide a data base to develop planning guidelines to determine the most cost-effective facility. Construction of several facilities would also increase experience with system performance to gain acceptance by engineers and the public. If it can be demonstrated that noncentral facilities are effective, regulatory agencies also may see the need for a change in policy.

ACKNOWLEDGMENTS

The assistance of Lester Forde, Research Assistant in the Department of Civil and Environmental Engineering, University of Wisconsin and Carl C. Crane, Inc., Consulting Engineers, in preparation of the Westboro facilities plan is appreciated. The legal research provided by David E. Stewart must also be recognized. The support given by the residents of Westboro was essential.

This study was supported by the Upper Great Lakes Regional Commission.

REFERENCES

1. Rezek, Henry, Meisenheimer and Gende, Inc., Libertyville, Illinois, unpublished data collected from U.S. 1970 Census (1975).
2. Beale, C. L. and G. V. Fuguitt. "Population Trends of Non-Metropolitan Cities and Villages in Subregions of the United States," CDE Working Paper 75-30, Center for Demography and Ecology, University of Wisconsin-Madison, Madison, Wisconsin (September 1975).
3. Smith, R. and R. G. Eilers. "Cost to the Consumer for Collection and Treatment of Wastewater," Water Pollution Control Research Series, 17090-07/70, U.S. Environmental Protection Agency, Washington, D.C. (July 1970).
4. Sloggett, G. R. and D. D. Badger. "Economics of Constructing and Operating Sewer Systems in Small Oklahoma Communities," Bulletin B-718, Agricultural Experiment Station, Oklahoma State University (April 1975).
5. Northwestern Wisconsin Regional Planning and Development Commission. "Model Facilities Plan for Three Unsewered Communities in Northwestern Wisconsin," Proposal for study submitted to the Wisconsin Department of Natural Resources (1974).
6. South Australia Department of Public Health. "Septic Tank Effluent

Drainage Schemes," Public Health Inspection Guide, No. 6, Norwood, South Australia (September 27, 1968).
7. Siegrist, R., M. Witt and W. C. Boyle. "The Characteristics of Rural Household Wastewater," *J. Env. Eng. Div. ASCE* 102(EE3):533-548 (June 1976).
8. Bowne, W. C. "Pressure Sewer Systems," Douglas County Engineer's Office, Roseburg, Oregon (May 1974).
9. Carcich, I. G., L. J. Hefling and R. P. Farrell. "Pressure Sewer Demonstration," *J. Env. Eng. Div. ASCE* 100(EE1) (February 1974).
10. Cliff, M. A. "Experience with Pressure Sewerage," *J. San. Eng. Div. ASCE* 94(SA5) (October 1968).
11. Committee on Public Health Activities of the Sanitary Engineering Division. "A Study of Sewage Collection and Disposal in Fringe Areas," Progress Report, *J. San. Eng. Div. ASCE*, SA2 (April 1958).
12. Bouma, J. "Unsaturated Flow During Soil Treatment of Septic Tank Effluent," *J. Env. Eng. Div. ASCE* 101(EE6) (December 1975).
13. Converse, J. C., R. J. Otis and J. Bouma. "Alternate Designs for Onsite Home Sewage Disposal," Small-Scale Waste Management Project, University of Wisconsin, Madison, Wisconsin (March 1976).
14. Sauer, D. K., W. C. Boyle and R. J. Otis. "Intermittent Sand Filtration of Household Wastewater Under Field Conditions," Proceedings, Illinois Private Sewage Disposal Symposium, Champaign, Illinois (September 29-October 1, 1975).
15. Sauer, D. K. "Treatment Systems for Surface Discharge of Onsite Wastewater," *Individual Onsite Wastewater Systems*, Nina McClelland, Ed., Proceedings of the 3rd National Conference, NSF (Ann Arbor, Mich.: Ann Arbor Science Publishers, 1977).
16. Otis, R. J. and D. E. Stewart. "Alternative Wastewater Facilities for Small Unsewered Communities in Rural America," Annual Report to the Upper Great Lakes Regional Commission, Small-Scale Waste Management Project, University of Wisconsin, Madison, Wisconsin (July 1976).
17. Otis, R. J., D. E. Stewart and L. Forde. "Alternative Wastewater Facilities for Rural Communities; A Case Study of Westboro, Wisconsin," Progress Report to the Upper Great Lakes Regional Commission, Small-Scale Waste Management Project, University of Wisconsin, Madison, Wisconsin (May 1975).
18. Douglas County Department of Public Works. "Sewerage Study for the Glide-Idleyld Park Area, Douglas County, Oregon," Oregon (December 1975).
19. Parrott, Ely and Hurt Consulting Engineers, Inc. "Sewerage Facilities Plan for Fountain Run, Kentucky," Lexington, Kentucky (July 1976).
20. Abney, J. L. "Integration of Onsite Disposal in a 201 Facilities Plan," *Individual Onsite Wastewater Systems*, Nina McClelland, Ed. Proceedings of the 3rd National Conference, NSF (Ann Arbor, Mich.: Ann Arbor Science Publishers, 1977).
21. Bowne, W. C. "The Collection Alternative: The Pressure Sewer System," *Individual Onsite Wastewater Systems*, Nina McClelland, Ed. Proceedings of the 3rd National Conference, NSF (Ann Arbor, Mich.: Ann Arbor Science Publishers, 1977).

22. Hinderman, D. W. Wisconsin Department of Natural Resources, Financial Aids Section, Madison, Wisconsin, personal communication (March 5, 1976).
23. Beck, J. and F. A. M. deHaan. "Phosphate Removal in Soil in Relation to Waste Disposal," *Proceedings of the International Conference on Land for Waste Management*, Ottawa, Canada (October 1973).
24. Green, K. M. and D. O. Cliver. "Removal of Virus from Septic Tank Effluent," *Home Sewage Disposal,* Proceedings of the National Home Sewage Disposal Symposium, ASAE Pub. Proc. 175 (December 1974).
25. McCoy, E. and W. A. Ziebell. "The Effects of Effluents on Groundwater: Bacteriological Aspects; Individual On-Site Wastewater Systems," *Proceedings of the Second National Conference,* National Sanitation Foundation, Ann Arbor, Michigan (November 1975).

27

INTEGRATION OF ONSITE DISPOSAL IN A 201 FACILITIES PLAN

Jack L. Abney

    Environmental Planner
    Parrott, Ely & Hurt Consulting Engineers
    Lexington, Kentucky 40502

OBJECTIVES

Fountain Run, Kentucky, is a small city that decided that reliance on individual sewage disposal was hindering its development. In 1976, a wastewater facilities plan[1] was prepared under a grant from the U.S. Environmental Protection Agency as provided in Section 201 of Public Law 92-500. The objectives of this plan were as follows:

1. Provide adequate public wastewater management to serve the needs of the community through 1995.
2. Comply with stream quality standards and other environmental regulations.
3. Minimize total 20-year costs for achieving the previous two objectives.
4. Develop a plan of implementation.
5. Assess environmental effects of various alternative systems that could meet the first three objectives.

CHARACTERISTICS OF THE PLANNING AREA

The planning area includes one incorporated city, Fountain Run, and about 3 square miles of unincorporated land, all in Monroe County, Kentucky. Most of the area is served by the Fountain Run Water District. No major water-using industries are located within the area.

The plan was prepared under the authority of the Water District, with the city cooperating.

The total population was 436 in 1975, with 318 living in the city. Lot sizes are fairly large, with the average city lot covering about one acre. About 130 residential and commercial occupied structures existed within the city limits in 1975.

Households and businesses all utilized onsite disposal of wastewaters in 1976. Most had septic tanks, but a few pit privies also were used. About 80% of existing wastewater sources were located on soils having good characteristics for subsurface disposal of wastewater. Major soil series include the Crider, Frederick and Trimble, which have USDA textural classifications ranging from silt loam to heavy silty clay loam.

The topography is rolling with some karst development. Underlying rocks are limestones and dolomite, with some interbedded shale.

WASTEWATER EFFLUENT STANDARDS

Any effluent discharging to a surface stream was required to meet fairly strict standards. Concentrations of key pollutants were not to exceed the following levels:

>   Five-day Biochemical Oxygen Demand: 10 mg/l
>   Suspended Solids: 15 mg/l
>   Ammonia Nitrogen: 1 mg/l
>   Dissolved Oxygen: 8 mg/l

ALTERNATIVE WASTEWATER MANAGEMENT SYSTEMS

In attempting to develop alternative systems, most of us are bound by our experiences, training and prejudices. One cannot usually consider an alternative that is not known or readily understood. Neither is a person likely to consider an alternative with which only negative experiences have been gained, unless forced to do so by regulatory or managerial edict.

Perhaps these are common reasons for not considering onsite disposal in engineering plans. But when a person has succeeded in breaking through the regulatory restraints against designing onsite disposal and has successfully designed systems on difficult sites, he is likely to consider this approach in future applications.

In 1965 the author was fortunate enough to be able to apply flexible design criteria for onsite disposal in a local health department in Indiana. Working from Federal Housing Administration studies of septic tank systems and with the aid of soil scientists and a geologist, he was able to develop a set of design criteria for onsite disposal that worked in that county very well. A further opportunity was gained in 1969, when he became associated with an environmental demonstration project in southeastern Kentucky. In that project several demonstrations of improved onsite disposal systems were installed on sites which could not be approved under the state plumbing code.

The Appalachian Project also had prepared several preliminary engineering plans for community sewerage systems. These plans included fairly detailed costs for sewer line construction which showed clearly the exorbitant cost of conventional sewers. Table I shows an analysis of these costs, updated by means of the EPA sewer construction cost index.

Table I. Appalachian Sewer Construction (Costs Updated to March 1976)

Type of Area	Users	Average Cost Per User	Maximum Cost Per User
1. Rural	596	$7,960	11,350
2. Rural	136	6,190	19,180
3. Rurban	2,025	5,970	26,280
4. Urban	73	5,730	12,420
5. Suburb	330	3,980	N/A
6. Urban	44	3,750	N/A
7. Suburb	335	3,470	5,240
All above		$5,860	

In some of the proposed service areas, sewer construction alone would cost more than the median annual family income of the persons served. When compared with the costs for the onsite disposal systems we had devised, sewers could not be economically justified in most of the areas studied in Appalachian Kentucky. However, no regulatory, financial or managerial system existed which would permit the effective utilization of "engineered" onsite disposal systems.

Therefore, the project could merely make recommendations for improvements in design of onsite disposal systems. These recommendations are contained in a report[2] published shortly before the demonstration was terminated.

In developing the Fountain Run plan, accepted federal policy was followed and, initially, only conventional sewers and central treatment were considered. Various treatment alternatives were examined, with simplicity of operation a primary goal. The final treatment process selected was a three-cell oxidation pond with land application of effluent. It was only after calculation of the average monthly bill that it was realized that the community probably could not afford such a system. Subsequent meetings with the local people confirmed that this conclusion was shared by community leaders.

With an assumed federal grant of 75% and a low-interest loan for most of the remaining 25%, the average monthly sewer bill would be over $17. With no grant, the average bill would be over $30. At the time, grants were available only for treatment, and so the higher figure would have been closer to reality.

Therefore, the consultants began to consider true alternatives to the safe, familiar, conventional sewers. The experimental sewer system installed at the Grady W. Taylor subdivision near Mt. Andrew, Alabama, served as initial inspiration for determining the cost of a similar system for Fountain Run. Further encouragement was given by recommendations developed by Paul Pate of the Birmingham, Alabama, Department of Health. Both capital and operating costs were projected to be lower for this "effluent sewer" system, as it was called. But the average bill would still be higher—about $13 per month.

It was then decided to divide the service area into smaller pieces and eliminate the central treatment facilities, while utilizing effluent sewers and subsurface disposal. This approach required a careful evaluation of the location of soils most suited to subsurface disposal and the identification of soil factors which might restrict their use for sewage disposal. Unit costs were developed for septic tanks, dosing devices, effluent sewers and disposal systems. Several trial-and-error combinations of users were tried before settling on a reasonably efficient arrangement. The final system consisted of 22 "community" systems having two or more users on shared disposal fields, plus 22 onsite disposal systems.

The cost for this "community subsurface disposal system" was significantly lower than the two previous systems. The average bill was estimated to be $7.30 per month, with 144 customers contributing. A further plus was the fact that an additional 24 customers were included.

As a final consideration, the cost for total onsite disposal with public management was analyzed. The same 144 customers were assumed to require replacement of their disposal system with new "engineered" systems. Standard absorption systems were estimated to cost $1200, while special designs required to overcome soil limitations were estimated to average $1800 each.

One design considered to have potential application was described in the above referenced Appalachian report. It consisted of an artificial "aquifer" of sand which was "recharged" by a dual sand filter. This concept is shown in Figure 1. It was believed that the filtered wastewater

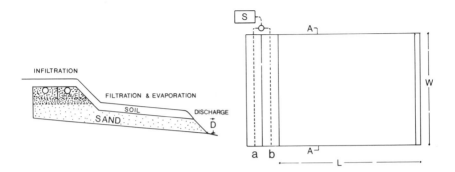

Figure 1. Septic tank-evaporation system.

would flow horizontally through the sand layer for evaporation from the overlying soil, given suitable climatic conditions. Wet weather and reduced insolation in winter would prevent evaporation. During these periods, filtered water would be discharged at the terminus of the sand bed. An empirical formula developed by G. T. Orlob was used to determine the horizontal filtering distance "L" required to reduce bacterial concentrations to safe levels. His formula, as reported by J. Timothy Winneberger, is:

$$L = 107 \, d \, V$$

where: L = distance of filtration
d = effective particle size of medium, in centimeters
v = velocity of liquid in feet per day

For an "average" washed sand with a d of 0.01 cm on a slope of 3%, L becomes 58 feet. The required cross-sectional area of the aquifer was determined by the formula:

$$A = \frac{q}{K_p \, S}$$

which is derived from the aquifer discharge formula as found in standard engineering texts. In that formula: K_p = coefficient of permeability in meinger units (gpd/ft^2); and S = slope of the hydraulic gradient. Local costs to construct such a system are estimated to be about $2000.

It was decided that most of the local soils could accept wastewater at very low rates of application from more conventional absorption trenches, resulting in lower system costs than the evaporation bed required. Such low-rate systems were estimated to cost $1800 each to construct, not including engineering costs.

The costs for the total onsite plan would be lower than costs for the community system. Average monthly billings would be about $5.70 with 75% federal assistance on construction costs.

ALTERNATIVES ANALYSIS

Table II summarizes the main features of these four alternate wastewater systems. Total construction costs, including engineering, for these alternatives is shown in Figure 2. Alternate A would require $524,400, B would require $367,500, C would require $340,200, and D would require $247,000. Conventional sewerage system A would cost more than twice as much as onsite disposal D, and 1.54 times as much as the community subsurface system C. This resulted even though a much greater cost for engineering and contingencies was included in C and D than in A. A rate of 20% was allowed in A, while 30% was allowed in C and D for engineering.

Table II. Alternative Systems

A. Central System	C. Decentralized System
Conventional sewers	Effluent sewers
Oxidation pond	Subsurface disposal
Infil-percolation	
B. Central System	D. Onsite System
Effluent sewers	Septic tank
Oxidation pond	Subsurface disposal
Infil-percolation	

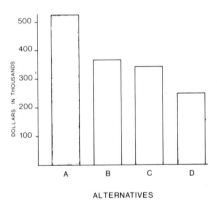

Figure 2. Total initial cost.

Annual funds required for operation, maintenance, billing and debt service are shown in Figure 3. Even though 20% fewer users are included in Alternate A, it would require 2.3 times as many annual dollars as D and 1.8 times as many as C.

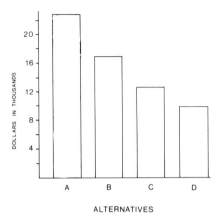

Figure 3. Total annual funds required.

Figure 4 illustrates the relationship in present worth for the four alternates. This comparison is the one mandated in the Section 201 planning guidelines. Present worth is a composite of initial capital, a lump sum to provide operation and maintenance for 20 years, and an allowance made for any salvage value at the end of the 20-year period. Relationships are similar to those in the preceding graphs, but not identical.

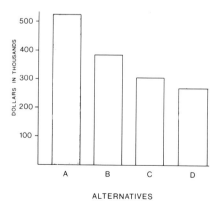

Figure 4. Present worth.

The economic comparison which matters most to the local citizen is the monthly bill for services. Due to the effect of federal funding and the varying service population, the relative difference in the four alternatives is greater than in any other comparison.

In all four alternatives, it was assumed that a federal grant for 75% of the initial cost would be obtained. In reality, this would be unlikely in any case.

But it could not be predicted with relative certainty how much of each alternative would be funded through a grant; and therefore equal outside funding was assumed. It was further assumed that a small "tap-in" fee would be charged each customer and the remainder of the local share would be borrowed over a 40-year period at 5% interest, the current FmHA loan terms. If no grant was available, the 75% portion would be financed locally through a greater loan and possibly a bond issuance.

The computed dollar amounts for these mean monthly bills were:

	With 75% Grant	With No Grant
Alternate A	$17.30	$37.80
Alternate B	12.80	27.60
Alternate C	7.30	17.30
Alternate D	5.70	12.90

These values are compared in Figure 5.

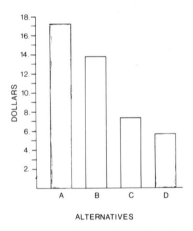

Figure 5. Average monthly bill.

This analysis indicates that Alternate A would cost the homeowner 3 times as much as D and 2.4 times as much as C with a grant. It is recognized that the long-term financing of 95% of a project, as is assumed in the "No Grant" column, is rather unrealistic.

It would seem very likely that most of the onsite users would be unwilling to pay the monthly cost in Alternate C without a grant.

SELECTION OF THE PREFERRED ALTERNATE

These data were presented to the officers of the water district, and a public meeting was held to explain the alternates to the affected citizens. It was decided by the District Board of Commissioners that Alternate C, the effluent sewer system with community subsurface disposal, would be the preferred alternate. Alternate D was not chosen because of a general feeling on the part of the citizens that no real advantage would be gained to justify the expenditure of $5.70 per month. Alternates A and B were rejected because of the high cost to the user.

DETAILS OF THE SELECTED ALTERNATE

The design of the selected systems is fairly simple. A septic tank and dosing tank would be placed at each user location. The effluent from the dosing tank will discharge into a plastic sewer of 4-inch inside diameter. Where the dosing tank must be located lower in elevation than the sewer, a sump pump will be used as described by Hindricks and Rees.[3] Otherwise a dosing siphon will be used to ensure scouring velocities near the lateral connection.

In a report by Otis and Stewart[4] effluent drains are described which have been used in South Australia since 1962, apparently with no need for such elaborate devices to provide a scouring velocity. But it would seem logical to expect a reduction in maintenance flushing of the sewers where intermittent dosing was provided. Effluent would be carried to the subsurface disposal fields via the plastic sewers. No manholes are proposed for these sewers, but cleanouts would be provided at intervals to allow flushing of lines should any sediment accumulate.

The preliminary design of the disposal fields is largely based on the work by Winneberger at Berkeley.[5,6] Field applications of the "narrow-trench" concept have proved successful in the author's experience in Jackson County, Indiana, and in Appalachian Kentucky. A comparison of the trench geometry required by Kentucky State Code with that recommended in the plan is shown in Figure 6. It may be readily seen that if the invert of the distribution pipe is considered the maximum design depth, then the narrow configuration provides an area per unit volume ratio of 2.33 times that of the standard configuration. Other calculations show that the total cost per useful square foot provided would be about one-half as much, using the narrower trench. Other design criteria are rather conservative, as outlined below:

Design Features for Subsurface Disposal

1. Application rate: 0.33 gpd/ft^2
2. Biennial use of alternate disposal trenches
3. Width-depth ratio of disposal trenches: 1 to 3
4. Design flow of 200 gpd/user
5. Intermittent dosing.

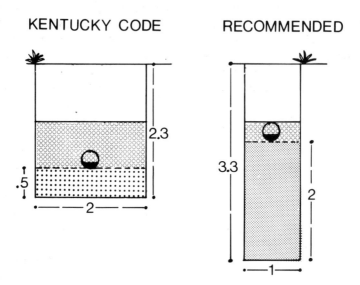

Figure 6. Disposal trenches.

In addition to the low application rate, two sets of trenches would be provided for use on a biennial cycle. Some persons have suggested that utilization of a biennial cycle should allow a reduction of up to 30% in the absorption surface area provided. However, we retained the more conservative approach.

In each disposal field, alternate trenches would be connected to a common header. This would allow a more diffuse application of effluent over the entire field. A design flow of 200 gpd per user may not seem very high, but the existing water consumption in Fountain Run is only 23 gallons per capita per day (gpcd) or about 70 gpd per customer. This preliminary design rate therefore provides for nearly three times as much flow as is presently occurring. Intermittent dosing of the disposal fields would be provided by either pumps or automatic siphons, depending on the size and topography of the field. This would help to provide uniform loading and avoid saturated flow through soil.

INDIVIDUAL ONSITE WASTEWATER SYSTEMS   287

In determining the optimum locations for disposal fields, available soil maps, topographic maps, aerial photographs and personal observations were utilized. Certain areas were eliminated due to the existence of soils with low permeability. Homes were grouped above available open land to try to achieve gravity flow to all disposal sites. Costs for effluent sewers were weighed against cost of disposal sites, convenience of maintenance and community acceptability. By a process of elimination, the total number of **multi-user sites** was reduced to 22, and 22 onsite systems in the built-up area were retained in the recommended plan. These latter users would receive a level of service equal to that provided the multi-user sites and would be charged at the same rate for services, if they choose to participate.

The pattern of septic tanks, effluent sewers and disposal fields obtained in the preliminary design is shown in Figure 7, which covers the central part of the incorporated city. Smaller subsystems as well as onsite systems would exist in adjacent areas but are not shown in this illustration.

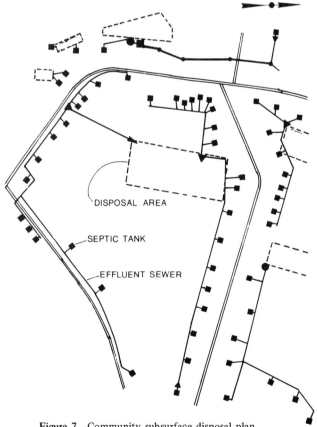

Figure 7. Community subsurface disposal plan.

Because of the uncertainties presented by several very small lots on the west side of the business district, a short length of conventional sewer leading to a central septic tank was proposed. If final design investigations show that septic tanks could be placed to serve these businesses, an effluent sewer may be recommended at that time.

Land on which the multi-user sites would be located would be owned by the water district. Land prices are low, due primarily to the low average income level and the lack of growth pressures. Accessibility to the onsite systems would be obtained by a utility easement, which it is assumed the homeowner would give in exchange for installing a new system that would be publicly owned and maintained.

Details of construction costs are shown in Table III. Unit prices were obtained from quotes by local contractors and recent bid tabulations for jobs. No significant allowances were made for possible quantity discounts. A summary of total materials and quantities provided in this Alternate is presented in Table IV.

Operating and maintenance requirements for the recommended system were more costly than might be expected. The system would contain 17 pumps of 1/3- to 2-horsepower size. Replacement units should be stocked in each size for rapid repair of malfunctioning pumps. Multi-user field dosing tanks would contain dual pumps for increased reliability.

The method used in computing total annual funds required is shown in Table V.

ENVIRONMENTAL EVALUATION

All of the four basic alternatives would appear to meet the effluent criteria and other environmental criteria of responsible regulatory agencies. As in most wastewater projects, the primary impacts are more readily determined than secondary impacts. The following discussion describes only the more significant environmental effects.

The following factors were considered in analyzing construction effects:

1. Erosion from sewer construction
2. Erosion from treatment and disposal sites
3. Stream-bank damage from sewer lines and treatment facilities
4. Aesthetic effects of excavation, etc.
5. Noise from construction equipment
6. Air quality effects from fugitive dust
7. The presence of sensitive ecosystems, unique plants, endangered species and archaeo-historic sites
8. Dislocation of individuals, businesses and governmental services
9. Employment.

Table III. Detailed Construction Costs for Preliminary Design
Community Subsurface Disposal System, Fountain Run, Kentucky

	Quantity	Unit	Unit Price	Total
Subsystem 1				
Onsite septic tanks	34	ea	$ 200.00	$ 6,800
Pumps, 1/3-hp w/tanks	1	ea	300.00	300
Small dosing siphons	33	ea	200.00	6,600
Effluent sewer, 4 in. dia	4,250	ft	4.00	17,000
Effluent sewer, 3 in. dia	790	ft	3.00	2,380
Main dosing tanks w/pumps	2	ea	1,200.00	2,400
Absorption trenches	10,200	lf	2.10	21,420
Land	2.5	ac	2,000.00	5,000
Subtotal				$61,900
Subsystem 2				
Onsite septic tanks	2	ea	200.00	400
Pumps, 1/3-hp w/tank	2	ea	300.00	600
Effluent sewer, 4 in. dia	450	ft	4.00	1,800
Gravity sewer, 8 in. dia, in place	950	ft	10.00	9,500
8 in. sewer fittings		ls		1,200
Manholes	5	ea	500.00	2,500
Main septic tank, 3000 gal	1	ea	750.00	750
Main dosing tank, w/pumps	1	ea	1,200.00	1,200
Absorption trenches	5,100	lf	2.10	10,710
Land	1.2	ac	2,000.00	2,400
Subtotal				$28,660
Subsystem 3				
Onsite septic tanks	12	ea	200.00	2,400
Small dosing siphons	12	ea	200.00	2,400
Effluent sewer, 4 in. dia	760	ft	4.00	3,040
Effluent sewer, 3 in. dia	240	ft	3.00	720
Main dosing tank, w/pumps	1	ls		1,200
Absorption trenches	3,600	lf	2.10	7,560
Land	0.8	ac	2,500.00	2,000
Subtotal				$19,320
Subsystem 4				
Onsite septic tanks	6	ea	200.00	1,200
Multi-user septic tanks	2	ea	300.00	600
Pump, 1/3-hp	4	ea	300.00	1,200
Small dosing siphons	4	ea	200.00	800
Effluent sewer, 4 in. dia	1,270	ft	4.00	5,080
Effluent sewer, 3 in. dia	200	ft	3.00	600
Main dosing tank, w/siphon		ls		550
Absorption trenches	3,300	lf	2.10	6,930
Land	0.8	ac	2,500.00	2,000
Subtotal				$18,960

Table III, Continued

	Quantity	Unit	Unit Price	Total
Subsystem 5				
Onsite septic tanks	7	ea	200.00	$ 1,400
Small dosing siphons	7	ea	200.00	1,400
Effluent sewer, 4 in. dia	1,080	ft	4.00	4,320
Effluent sewer, 3 in. dia	140	ft	3.00	420
Main dosing tank, w/siphon		ls		550
Absorption trenches	2,100	lf	2.25	4,725
Land	0.5	ac	3,000.00	1,500
Subtotal				$14,315
Subsystem 6				
Onsite septic tanks	3	ea	200.00	600
Multi-user septic tank	1	ea	300.00	300
Small dosing siphons	4	ea	200.00	800
Effluent sewer, 4 in. dia	720	ft	4.00	2,880
Effluent sewer, 3 in. dia	100	ft	3.00	300
Main dosing tank, w/siphon		ls		550
Absorption trenches	1,800	lf	2.25	4,050
Land	0.75	ac	2,500.00	1,875
Subtotal				$11,355
Subsystem 7				
Multi-user septic tanks	2	ea	300.00	600
Small dosing siphons	2	ea	200.00	400
Effluent sewer, 4 in. dia	230	ft	4.00	920
Absorption trenches	1,200	ft	2.25	2,700
Land	0.5	ac	3,000.00	1,500
Subtotal				$ 6,120
Subsystem 8				
Onsite septic tanks	3	ea	200.00	600
Small dosing siphons	2	ea	200.00	400
1/3-hp pump w/tank	1	ea	300.00	300
1/2-hp pump w/tank	1	ea	450.00	450
Effluent sewer, 4 in. dia	500	ft	4.00	2,000
Effluent sewer, 2 in. dia	100	ft	3.00	300
Absorption trenches	900	lf	2.25	2,025
Land	0.6	ac	2,500.00	1,500
Subtotal				$ 7,575
Subsystem 9				
Onsite septic tanks	3	ea	200.00	600
Effluent sewers, 4 in. dia	300	ft	4.00	1,200
Main dosing tank w/siphon		ls		400
Absorption trenches	900	lf	2.25	2,025
Land	0.33	ac	3,000.00	1,000
Subtotal				$ 5,325

Table III, Continued

	Quantity	Unit	Unit Price	Total
Subsystem 10				
Onsite septic tanks	3	ea	200.00	$ 600
Small dosing siphons	2	ea	200.00	400
Effluent sewer, 4 in. dia	350	ft	4.00	1,400
Effluent sewer, 3 in. dia	100	ft	3.00	300
Main dosing tank, w/siphon		ls		400
Absorption trenches	900	lf	2.25	2,025
Land	0.5	ac	2,500.00	1,250
Subtotal				$ 6,375
Subsystem 11				
Onsite septic tanks	3	ea	200.00	600
Small dosing siphons	2	ea	200.00	400
Effluent sewer, 4 in. dia	400	ft	4.00	1,600
Effluent sewer, 3 in. dia	50	ft	3.50	175
Main dosing tank, w/siphon		ls		400
Absorption trenches	900	lf	2.25	2,025
Land	0.33	ac	3,000.00	1,000
Subtotal				$ 6,200
2-Unit Disposal Systems (Sites 12 through 22)				
Onsite septic tanks	14	ea	200.00	2,800
Multi-user septic tanks	4	ea	300.00	900
Small dosing siphons	14	ea	200.00	2,800
1/3-hp pump & tank	1	ea	300.00	300
Effluent sewer, 4 in. dia.	1,220	ft	4.00	4,880
Absorption trenches	6,600	lf	2.25	14,850
Land cost	1.8	ac	3,000.00	5,400
Subtotal				$31,930
Individual Disposal Systems				
Onsite septic tanks	22	ea	200.00	4,400
Small dosing siphons	22	ea	200.00	4,400
Absorption trenches	6,600	lf	2.25	14,850
Subtotal				$23,650
Total treatment & disposal costs				$241,685
Sludge pump, soil injector and truck				20,000
Grand Total				$261,685

Table IV. Summary of System Components

122 septic tanks
13,250 linear feet of effluent sewer
960 linear feet of 8 in. sanitary sewer
104 small dosing siphons
9 small effluent pumps
4 main dosing tanks with pumps
6 main dosing tanks with siphons
44,100 linear feet of absorption trenches at 44 sites
10.6 acres of land
1 set sludge pump and soil injection equipment

Table V. Alternative C Annual Fund Requirements

Operation and maintenance		$ 6,100	
Office and billing expense		1,000	
Subtotal			$ 7,100
Debt Service			
Construction cost		$261,685	
Engineering, legal & contingencies @ 30%		78,506	
Total initial cost		$340,191	
Less grant (75%)	(-)	255,143	
Local share		85,048	
Less tap-on fees @ $50 ea (130)	(-)	6,500	
Net debt amount		$ 78,548	
(Assume 40-yr loan @ 5% use capital Recovery factor of 0.05828)			
Average annual principal & interest		$ 4,578	
Surplus for reserves at 20%	(+)	915	
Total debt service funds			5,493
Total annual funds required			$12,593

Erosion

Erosion was estimated for each alternative by assuming a uniform soil erodibility (K factor) for the soil (the dominant soil type) and uniform erosion control practices (mulching) and estimating the steepness of the affected area by use of a topographic map. The Universal Soil-Loss Equation was applied to these assumptions and the total annual soil loss was adjusted to the estimated time of construction exposure. The results of these calculations are summarized below.

Alternative	Soil Loss, Tons
A	30
B	25
C	6
D	2

From these data it may be predicated that the conventional gravity sewer system and lagoon with disposal in an infiltration basin (Alternative A) would create the greatest soil loss. The least soil loss would be created by Alternative D, using onsite disposal, and Alternative C would create slight higher losses than D because of the effluent sewers being provided. Because all such losses would be distributed over a fairly large area in a "non-sensitive" environment, no significant adverse impact would be anticipated.

Stream-Bank Damage

Damage to stream banks in the form of earth cuts and fills would be experienced in Alternatives A and B as a result of the construction of a 2-acre lagoon in the bed of a stream. This construction would require the diversion of the intermittent stream around one side of the lagoon. Additional damage could occur from construction of sewer lines crossing streams in Alternatives A and B. Alternatives C and D would not cause such damage since no major construction is proposed in any stream.

Noise

Because larger construction equipment generally produces greater noise levels, Alternatives A and B would tend to produce greater significant noise impact than Alternatives C and D. However, the most noise would be produced by bulldozers constructing the lagoon, and the lagoon site is located more than 500 feet from the nearest residence.

Sewer line construction in an existing community often produces noise levels which exceed the EPA criteria for noise. Because the construction of effluent sewers, as in Alternatives B and C, would be done with smaller equipment, fewer excessive noise incidents would be expected. The total lack of pavement crossings in Alternative D would indicate that this alternative would have least adverse noise impact.

Other Impacts from Construction

All other potential impacts from construction were considered to be insignificant. No rare or endangered species, sensitive ecosystems or historic sites would be adversely affected by any alternative considered feasible.

Groundwater Effects

All of the final alternatives utilized some form of disposal to the soil. Construction of soil conditions and the hydrogeology of the area have shown that the possibility of groundwater contamination by the proposed facilities is remote. All soils considered for disposal are fine textured and moderately well drained. They are considered to have a large capacity for absorption of ammonia nitrogen, nitrate and phosphorus. No high groundwater conditions were evident in any disposal area.

Concern is often expressed in engineering reports about nitrate contamination of groundwater below septic-tank effluent disposal fields. An attempt was made by Rajagopal et al.[6] to relate groundwater quality to septic tank densities in an area with sandy soils and fairly high water table. In 123 samples, only nitrates were found to approach or exceed U.S. PHS limits for drinking water. Only six samples had concentrations in excess of the standard, and these were apparently caused by fertilization of cherry orchards, not by septic tanks. Where no orchards were nearby, nitrate apparently did not exceed 2 ppm (as NO_3-N) average concentration.

More detailed consideration of the location of any existing wells will be made in the Step 2 (design) process. Nearly all persons in the area of concern are customers of the water district, but a few private wells may still exist and, if so, adequate separation distances from disposal sites must be provided or the wells should be abandoned and sealed.

The potential for overflow of partially treated wastewater in the effluent sewer system is probably much less than the potential for overflow of raw wastewater in the conventional system. This is due in part to the provision of onsite storage of several hours capacity in the dosing tanks and septic tanks. A typical 1000-gallon septic tank would have a reserve storage capacity of about 100 gallons with a rise of 6 inches in liquid level. This would equal about 12 hours of average flow, which should be sufficient time to complete most repairs or replace failed pumps.

In addition, hydraulic overloads from infiltration and inflow appear to be much more likely with conventional sewers than with effluent sewers, due to the relative integrity of joints and the presence of manholes in the conventional system. Investigations of infiltration and inflow in existing sewer systems have demonstrated that untreated discharges were common in all systems.

Further protection against accidental overflow in effluent sewers could be provided by small emergency sand filters located adjacent to disposal field pumps, or by emergency subsurface disposal trenches.

SUMMARY OF CONSTRUCTION AND OPERATING EFFECTS

Each of the effects described above has been given a numerical rating and the ratings added to give a total for ranking purposes. This process is shown in Table VI. The results would indicate that the five alternatives do not exhibit a very wide spread in relative environmental impact. The total rankings, in ascending order of possible negative impact, were:

1. Alternative D = 29
2. Alternative C = 30
3. Alternative B = 37
4. Alternative A = 40

SECONDARY IMPACTS

Conventional gravity sewers often are considered to stimulate growth and encourage new industry to move to an area where excess capacity exists in a sewerage system. Of course, this is dependent on many other factors as well, such as availability of general and skilled labor, transportation facilities and distance to markets. Nevertheless, it would appear likely that conventional gravity sewers as considered in Alternative A would tend to cause more development and, therefore, create a potential for greater secondary impact than Alternatives B, C and D.

EVALUATION OF IMPLEMENTATION

Based on the consultant's understanding of the powers of water districts, any of the alternatives could legally be implemented by the district. Alternatives C and D are apparently unique proposals in Kentucky and for that reason, may entail more original thought and careful evaluation for successful implementation.

On the other hand, the conventional sewer with central treatment would require such a large expenditure of local funds, even with federal

Table VI. Summary of Environmental Assessment for Final Alternatives

Negative Impacts	A-1	A-2	B	C	D
Primary					
Erosion	3	4	3	1	1
Stream damage	3	5	4	1	1
Aesthetics	3	4	3	3	3
Noise from construction	6	6	4	3	3
Odor	1	1	1	1	1
Fugitive dust	5	5	4	3	2
Air pollution	1	1	1	1	1
Natural communities	1	1	1	1	1
Sensitive areas	1	1	1	1	1
Scientific & cultural resources	1	1	1	1	1
Dislocation	1	1	1	2	2
Employment	1	1	1	1	1
Groundwater quality	2	2	2	2	2
Surface water quality	1	1	1	1	1
Energy consumption	3	1	1	1	1
Noise from operation	2	1	1	1	1
Secondary					
Development	3	3	3	2	2
Pollution from development	2	2	2	2	2
Damage to ecosystems	1	1	1	1	1
Damage to sensitive areas	1	1	1	1	1
Totals	40	43	37	30	29

assistance, that opposition from potential customers may be even greater than anticipated. Alternatives A and B may also require a trained operator, or at least require considerably more manpower than the other alternatives, which would be a disadvantage.

Alternative D would seem particularly difficult to implement from the standpoint of the 20% of homes located on soils of low permeability. As pointed out previously, that alternative could involve much higher costs for design and construction of the systems located in poor soils than was used to determine relative present worth. From this standpoint, Alternative D is not recommended.

In perspective, none of the alternatives had any overwhelming advantage for implementation. Further consideration of implementation is contained in the Facilities Plan.

PUBLIC PARTICIPATION

A notice of public hearing for discussion of the environmental inventory and alternatives developed in the plan was published in a local newpaper. About 15 community leaders attended the hearing.

Discussion during the meeting centered around Alternative C, the community subsurface disposal system. Alternative A, conventional sewers and central treatment, was considered too expensive by all participants. Even Alternative B, the effluent sewer system with central treatment, was considered too expensive for local income levels. Several participants mentioned the fact that a significant portion (local estimates were 30%) of the population was living on retirement income and Social Security.

It was pointed out that the element of risk of "failure" may be higher in Alternative C than with conventional sewers, because of the complexity of soils and relative sensitivity to errors, but that any failure would probably affect only a few persons and would be correctable. The importance of the central management concept to correcting problems was explained.

To those attending who had no immediate problem with their individual disposal systems, even the expenditure of $7 per month seemed to be little justified when the discussion was commenced. An objection to Alternative C was that it might not attract new industry in the manner hoped for by some citizens. Some questioned whether as many persons would "sign up" for services as had been projected, and this led to a discussion of the possible mandating of subscriptions by health authorities or city ordinance. An opinion of the State Attorney General advised that water district commissioners would have legal authority to require use of a sewer system.

The participants largely agreed that the community disposal system would be a more desirable improvement and that Alternative C would probably not cost any more than maintaining and replacing existing septic tank systems. Several persons mentioned neighbors and business places where septic tank failures were known but have not been corrected.

Because the majority favored Alternative C, subsurface disposal, as a result of the lower cost and simplicity of operation, the chairman instructed the consultants to proceed with Alternative C as the preferred alternate.

CONCLUSIONS

The community subsurface disposal concept favored in this plan is not a new concept; but it has had little, if any, application. To the best knowledge of the author, no demonstration has included the mix of septic tanks, effluent sewers, community subsurface disposal and onsite disposal recommended in the plan. Because the overall concept is somewhat new and unfamiliar to the federal funding agencies, the possibility of substantial federal assistance is unknown. On the Kentucky State Priority Ranking, the project is listed as 240th out of 241. This low rank is due primarily to the lack of recognized wastewater discharges in the local area. Malfunctioning septic tank systems are not included in the weighting system for determination of need.

The low ranking not only delays funding of the project, but also delays approval of a project. Personnel at the EPA regional office have stated that the Fountain Run plan would not be reviewed for approval until higher ranking projects had been reviewed, and no timetable for such review was available. The most recent advice from the state office is that funding of this project is at least 10 years in the future, assuming funding levels do not increase. Therefore, other sources of funding are being investigated.

More immediate application of the effluent sewer concept could be made by financing only the more central subsystems through a long-term loan. This might provide a base from which the district could expand services as more customers become convinced of the advantages of public management of their household wastewater.

In discussing the proposed system with persons in various positions, from citizen to regulator, it seemed that most persons are initially prejudiced against all these key elements of the concept. Retaining septic tanks at the individual wastewater sources seems to violate what most sanitary engineers and citizens feel is right—that all wastes should be carried away from the point of generation as quickly as possible. Similarly, it violates common practice to specify a sewer with as small as a 4-inch diameter, when in some local jurisdictions 8-inch sewers are laid right up to the house foundation. And the history of subsurface disposal of wastewater has been so filled with negative experiences that regulatory officials sometimes are unable to give this alternative serious consideration.

So it appears that the chief obstacle to implementation of such systems in Fountain Run and many other places is not technical or financial, but psychological and social. But we are beginning to honestly face the financial and environmental costs of the conventional sewer and central treatment systems. As we do so, we may be able to accept the advantages of community subsurface disposal, and onsite disposal, and utilize these methods to reduce the costs of wastewater management in suburban and rural areas.

REFERENCES

1. "Sewerage Facilities Plan—Fountain Run, Kentucky," Parrott, Ely and Hurt Consulting Engineers, Inc., Lexington, Kentucky (July 1976).
2. Abney, J. L. "On-Site Sewage Disposal Systems—Technical Considerations and Recommended Design Approaches," Appalachian Environmental Demonstration Project, Kentucky State Department for Natural Resources and Environmental Protection, Corbin, Kentucky (June 1973).
3. Hendricks, G. F. and S. M. Rees. "Economical Residential Pressure Sewer System with No Effluent," SIECO, Inc., Columbus, Indiana EPA-600/2-75-072 (December 1975).
4. Otis, R. J. and D. E. Stewart. "Alternative Wastewater Facilities for Small Unsewered Communities in Rural America," Small-Scale Waste Management Demonstration, Phase III, Annual Report, University of Wisconsin, Madison (July 1976).
5. Winneberger, J. T. H. and P. H. McGauhey. "A Study of Methods of Preventing Failure of Septic-Tank Percolation Systems," SERL Report No. 65-17, Sanitary Engineering Research Laboratory, University of California, Berkeley (1965).
6. Rajagopal, R., R. L. Patterson, R. P. Canale and M. J. Armstrong. "Water Quality and Economic Criteria for Rural Wastewater and Water Supply Systems," *J. Water Pollution Control Fed.* 47(7) (July 1975).

28

INNOVATION IN WASTEWATER TECHNOLOGY: THE CHALLENGE OF THE 1980s

Kenneth C. Pearson

    Interlink Life Support Systems, Inc.
    Costa Mesa, California 92626

INTRODUCTION

The logic for onsite or decentralized wastewater systems is now clearly established in terms of conservation of our water resources, energy resources and the environment. The thrust of technological innovation today recognizes that water reduction, or its removal from the traditional role of sewer transportation, is a key design criterion. The objective of this paper is to review the three levels of resistance to the widespread implementation of onsite devices and relate these to the emerging onsite technologies. A better understanding of the interface between this new technology and the impediment levels will aid in formulating a new climate of acceptance at all levels of government and involved bureaucracies.

RESISTANCE LEVELS SUMMARIZED

We recognize three clearly defined levels of resistance to the reordering of our wastewater management priorities from central systems to onsite systems.

Based on consumer acceptance criteria, technology has to provide acceptable alternatives which meet aesthetic and functional performance levels. Provided that public health criteria are met, the second level of resistance is clearly the skepticism of a large proportion of public health bodies across the nation. This skepticism is due in part to prior association with inferior technology, unsuitable technologies assimilated to domestic situations and, in some cases, an unwillingness to accept new technological advances.

In stating the precepts for the third level of this resistance review, we state the use of sanitation as an urban planning weapon to aid political bodies in restricting the growth of their areas. The conflicting interests of developers and landowners, civic groups and local county regulatory authorities seldom blend in harmonious accord.

In many cases the resolution to these political issues has been achieved by adopting restricted zoning laws based on sanitation issues. Technically, many of these cases have been justified, but a true climate for change will not eventuate if new technologies are denied acceptance because of political factors.

With this overview of the problem, I would like to suggest some initiatives which, if acted upon, would provide the necessary criteria for the three levels of acceptance which are required in this area.

CONSUMER RESISTANCE

As with any embryo or growth industry, the consumer is often exposed to new technologies which have been reordered or restructured from existing technology. In the domestic sector, many companies have attempted to market systems which were never designed for the typical urban home. I exclude from these comments the holiday home market. In this regard public health and building codes should be the barrier between marketer and consumer. The criterion for consumer acceptance is as undefinable as the vaguaries of the consumer. However, I list certain criteria which we believe are realistically necessary in light of three to four generations of Pavlovian potty training. In introducing these criteria, an assumption is made that acceptable alternative technologies will reduce or eliminate the use of water in the toilet.

1. Acceptable alternative systems should cater to the out-of-sight, out-of-mind syndrome. The implication is a flushing or evacuation system akin to that provided by the water flush systems.

2. Totally automated systems are a necessity because of consumer apathy in the maintenance of toilet systems. We believe that hands-off systems are less likely to fail than those relying on regular consumer involvement in the functional modes.

3. Energy use must be low in light of existing energy realities.

4. Pollution aspects are now of paramount importance. The design criteria should strive for no discharge of human waste. This ideal is

recognized by several systems developers who have adopted split-stream systems where human waste containment or disposal is offered onsite. Management of greywater, although a problem, can achieve the nil waste discharge goal by meeting surface or waterway discharge requirements.

5. One final design objective must be recognition of economic restraints. No alternative system is going to find widespread acceptance unless it can meet a realistic cost-effectiveness goal. My own company's research places a bracket figure of $2000 to $3500 per domestic unit as an acceptable cost to the developer today, provided such a system manages the total household effluent flow.

PUBLIC HEALTH ASPECTS

New technologies must prove their worth in our society. Public health interface is vital to protect the consumer and the community. What is the key to open the door for open exchange between the manufacturer and public health officials? This contentious area has never been resolved. However, certain logical steps are open to both parties, if they wish to avail themselves thereof.

1. As a first step the manufacturer must be able to demonstrate that all relevant public health aspects have been satisfied. Traditionally, manufacturers have relied on four approaches to demonstrate system feasibility:

- Monitored programs in which local public health officials are invited to monitor provisionally approved systems.
- Submission of independent commercial laboratory reports.
- National Sanitation Foundation (NSF) certification programs.
- Data accumulated from systems used in the government, military or public sectors.

As every manufacturer here is aware, there is no nationally recognized route to total acceptance of a new system.

2. Public health officials can rightly be concerned that no single existing testing program outlined above has, in the past, satisfied all reasonable criteria for domestic situations. For example, with the exception of a monitored program in a specific county:

- The laboratory cannot always approximate domestic realities in use and abuse.
- Existing NSF standards have not always addressed the specific issues involved with domestic situations.
- Domestic reorientation of systems tested in the public, military or government sectors often are unrealistic because onsite

routine maintenance is available to an extent unlikely to be encountered in urban situations. Domestic realities are seldom approximated in this sector and, conversely, public use systems are designed for far greater stresses and abuse than those normally prevalent in the family home situation.

Each one of the traditional proving grounds could, with modification, become suitable and adaptable to most public health requirements. My own opinion as a manufacturer is that NSF provides the most acceptable, centralized program for across-the-board acceptance, provided that certain issues which are now being addressed come to pass.

1. The adoption of a fair but strict standard which is recognized by public health as satisfying its concerns. A joint task committee has recently completed a standard which can meet this criterion.

2. Where appropriate, the testing of systems should be in actual domestic use conditions, a goal which NSF recognizes and is responsive to.

3. By addressing the greywater problem, which if left unresolved will continue to negate a lot of interesting, innovative research now being centered on human waste management and disposal technologies, but which cannot stand alone in the domestic market.

POLITICAL ASPECTS

As an Australian who has had two years exposure to the realities of politics in this country, I candidly state that there will never be a permanent, logical solution to the political factors outlined earlier. One can isolate the educational role which must be borne by some group to whom the politicians are responsive.

There is little doubt that local politics and their regulation of property development and the environment are the pertinent issues involved here. Federal mandates are inappropriate at local levels. The federal executive controls the purse strings, and it is encumbent upon this body to initiate funding programs responsive to alternative systems.

I believe that the state level is where the educational process should begin, with appropriate national guidelines issued by the Environmental Protection Agency (EPA). Water conservation, environmental control and energy aspects are of concern at all government levels.

The mandate to improve the climate for new onsite technologies, which truly address the issues discussed at this conference, probably and realistically belong at the state level. The states in turn can rightfully look to EPA to provide not only policy guidelines and funding, but impetus to programs which can and will improve the climate for change in an area which has seen little effective change since Sir Thomas Crapper's invention.

With state enlightenment there inevitably will follow the necessary local government reaction. Certainly the prospect of demonstrating technical innovation at 51 state levels is substantially less a burden to a manufacturer than the task involved with dealing with over 3500 county and city bureaucracies.

CONCLUSION

To place these remarks in perspective, I would outline the challenge of the 1980s as it now confronts the vested interest parties. The initial onus rests with the industry to provide viable, acceptable alternatives, geared specifically to the domestic sector. This technology must pass the scrutiny of any essential service which is no easy feat in the light of all requirements today. Has the technology arrived yet? I can state categorically that it has. Has it met all the criteria listed in this paper? My review of the industry shows me that a small handful of corporate efforts are now within 1 to 3 years of achieving all of the stated criteria. The initial commercial success of the first systems will ensure that competition and subsequent rapid improvement will surely follow to create a new multibillion dollar industry. Will public health resistance to change remain forever an obstacle to this industry? I think not. This key lies in providing realistic demonstrability of feasible alternatives which address all public health concerns. Effective communication, fair codes and common sense will then break down resistance level 2.

Finally, will the political issues remain the fluid, pressure group-susceptible factor they often are today? Ultimately, I would think not. The laws of diminishing resources, new public awareness of the environment and the subsequent action-reaction at federal, state and local levels will ultimately show its response to the tide of public opinion. The society that placed man on the moon needs only the motivation to finally start cleaning up its own backyard.

29

POLLUTION CONTROL—
PATHWAY TO PERFECTION OR PERDITION

Dale A. Carlson

>Dean, College of Engineering
>University of Washington
>Seattle, Washington 98195

The disposal of wastes receives attention when there are disturbances to the environment or inconveniences to man. The disposal systems thus far contrived to alleviate waste problems actually involve only a redistribution of materials originally extracted from the environment.

Disposal involves consideration of the stability of compounds being returned to the environment; the potential for movement of waste components beyond the disposal site; the possibility for disruptive influences at the disposal area and its environs; and the possibility for beneficial uses of waste fractions.

The basic problems in waste disposal are to remove the energy insofar as possible from waste streams prior to release to the environment, and to separate solids, liquids and gases. Once the negentropy of the wastes is reduced, the wastes should be transported to locations such that energy is not immediately available to incorporate waste nutrients into nuisance accumulations, such as excess algae.

Separation of water from solids traditionally has been divided into: 1) primary systems which provide physical separation via gravity and screening, sometimes including chemicals to enhance removal; 2) secondary systems using mixed biological populations to stabilize and coalesce organics; and 3) tertiary systems which include chemical processes. The trend has been to add on increased levels of treatment to meet higher effluent standards. At the same time, the collection of wastewaters in larger and larger plants has progressed to the point at which feasibility becomes limited with distance.

It is possible to increase treatment to the point at which effluents approach drinking water standards. Such high levels of treatment require energy and chemicals, the production and transport of which involve the concomitant production of wastes. There is thus a treatment level at which the amount of pollutants removed is exceeded by the amount of pollution generated in chemical and energy production and transport. The crossover is somewhere over 90% removal of solids from domestic wastes. The data in Table I illustrate the crossover concept.

Table I. Components from Treatment Systems

	Primary Treatment	Secondary Treatment	Treatment to Drinking Water Standards
Tons of solids removed per year	2,500	1,500	4,700
Tons of chemicals and resources used per year	4,000	2,400	40,000
Energy used in millions of kwhr per year	2.6	4.2	14
Tons of pollutants generated per year	3,000	1,000	19,000
Billions of Btus generated per year	17	25	90
Tons of pollutant generated/tons pollutant removed	1.2	0.7	4.0
Tons resources used/tons pollutant removed	1.6	1.6	8.5

Source: Ling, J. T., *Industrial Water Engineering*, January-February 1973, pp. 14-17.

The treatment processes do alleviate aggravations caused by the redistribution of wastes back to the environment. The conventional processes do not, however, prevent the growing dimensions of some long-term problems, which must be addressed and which in their diminution require modification to the treatment trains. Among these long-term problems are resource depletion with resultant higher energy and cost requirements; the accumulation of resources in the wrong places in dispersed, rather than concentrated, form; and loss of soils from food production areas at exorbitant rates.

Prior treatment or preventative treatment are mechanisms which can mitigate pollution control requirements. Such mechanisms involve process quality control, conservation, adequate design, and control of system loading sequences. These systems may be especially important where toxic or inhibitory materials can be prevented from entering a waste stream, thereby enhancing the potential for land disposal of treated effluents.

The major problem for requiring treatment can be the sheer quantity going back to the environment; however, the rate of transport of wastes may be more important than quantity. The transport routes of waste streams may be analogous to river flow or flow in flexible tube pipelines. The flooding of rivers disrupts all normal river use, just as overloading of waste transport systems can disrupt or destroy biological recipient populalations. Likewise, the rupture of overloaded flexible tubes with excess pressure destroys the transport system completely.

While all treatment systems involve trade-offs and there is no perfect treatment system, it is necessary to assure that rate and quantity are considered throughout the disposal process. Under ideal conditions, the use of soil systems for disposal should, for example, allow the escape of relatively pure water to the water table; retain nutrients at levels accessible for uptake by the proper flora; keep nitrates out of water systems and heavy metals out of plants; return organics to useful soil tilth; avoid odor production and matting, while maintaining the normal growth in the receiving area.

Treatment and disposal of wastes involves ultimately the well-being of man and the maintenance of his life support systems. Because food production is becoming more important as the "American Age of Surplus" draws to a close, the coming "Age of Critical Resource Shortages" will dictate that a major consideration be the optimization of resource use and resource disposal and return to the land. While the return of organics to the soil seems a logical salvage process, the incorporation of metals into the waste stream poses long-term storage and uptake problems, both for plants and for man.

If toxic metals cannot be recaptured prior to land disposal, the term design period should be considered as the storage period of the toxic metals in man, as well as the possible genetic carryover period to future generations. The engineering design of land treatment must then include the review of the pathway of metals from mining to use to redistribution, uptake in crops, and uptake and storage in man and animals.

What remains to be learned about disposal and environmental impact may be more important than what is already known. For example, it was thought that landfill leachates did not require chlorination because they were devoid of interior organisms. Yet, recent research indicates that, while the concentrated leachate stream does not reveal large enteric populations, mere dilution and aeration over a day are sufficient to provide significant enteric organism viability in the diluted leachate. Thus, one must still be concerned with the need for leachate disinfection prior to its entrance into aerobic streams.

Continued research by knowledgeable teams in such areas as biochemistry of soils and humans, geology, physiology, enzyme chemistry, soil physics, and engineering will be needed to fully utilize land disposal systems. In these efforts engineering is the coordinative role of applying science and art to the servicing of man's needs. The application of engineering in industry then must be useful to industry, but that use is subservient to alleviating the hurts of man.

Industry, engineering and government all should be intended to enhance life and civilization over the long run. In dealing with the complexities of living systems, it is necessary that we serve each other for the betterment of society.

The choice, then, is ours to chance the road to perdition by self-serving, short-term profits and neglect of the trade-offs available, or to strive for perfection by taking a longer-range perspective of what is better use of the environment, not only for this generation, but also for succeeding generations of man.

30
AS WE NOW LOOK AHEAD

Robert M. Brown
    President, National Sanitation Foundation
    Ann Arbor, Michigan 48105

    It has been a rewarding experience for us at NSF to have sponsored, together with our colleagues in the Technology Transfer Program of EPA, this Third National Conference on Onsite Wastewater Systems. There can be no doubt, as we look back to the first conference held just two years ago, that there has been real progress in determining the appropriate role of onsite systems in meeting today's and tomorrow's needs. This is reflected in the growing numbers of interests working effectively together to define basic issues. Enthusiasm replaces discouragement—as a mechanism for interfacing between parties at interest seems to be emerging in a way that most find acceptable.
    In particular, I believe we are all aware of accomplishment in defining the problem of experience in use of onsite methods of wastewater handling. We are similarly aware of progress which has been made in advancing and supplying technology in refining systems and their applications. At the same time we mark a need for more definitive data on potential epidemiological effects of onsite system discharges to surface or groundwaters. There appears to be more willingness, with perhaps a growing confidence, to move the attack into the economic, social and political sectors where involvement and acceptance are essential to the ultimate formulation of public policy.
    Without question we can note the building of a momentum in activity to find and supply appropriate means for dealing with wastewater disposal problems where conventional sewerage methods fall short. Perhaps our greatest challenge at this point in time is to maintain such momentum and, if at all possible, to accelerate it. The three national conferences

which NSF has hosted have reflected a build-up in action and determination involving both U.S. and Canadian interests clearly beyond enthusiasm evident among conference participants.

From all indications, the Third National Conference fulfilled its objectives. As we now look ahead, our plan for the future should be developed in pragmatic terms. This is to say that we know where we are, we know with reasonable confidence where we want to be, and we are now prepared to lay out the course we think best to get there. It would be entirely appropriate, to be sure, to define a few alternative routes to be followed to maintain momentum in the event *to-be-anticipated* road blocks materialize. Let me suggest some next steps.

It would be broadly beneficial to undertake, forthwith, an objective overview of where we are with respect to the use of onsite systems pursuant to the issues dealt with at this third conference. From presentations and discussions, there appears to be an emerging new philosophy as to the acceptance and use of individual onsite systems. If this is indeed real, it should be possible to define and assess progress in this direction which has been observed in the two-year period since the First National Conference in 1974. This progress should be related to advances in *technology, administrative management* and *societal acceptability.*

There seem to be areas of knowledge relating to human health or environmental ecology where additional epidemiological data would provide greater assurance for proceeding with plans for surface discharge of effluents, or discharges reaching groundwater. A collection and analysis of all currently existing data could lead to a determination of whether additional data is required. The objective is to know that adequate epidemiological intelligence assures the public health safety of a new pattern of onsite system usage.

There is need to achieve a better understanding of the economic, social and political factors which have an influence upon acceptance and use of onsite systems. With such an understanding, better skills can be developed in leading—or at least constructively influencing—societal decision-making processes. In many instances factors of fear and emotionalism seem to have a disproportionate influence upon public attitude and administrative decision-making. Urgently needed is a definition of the approach to be taken to develop an informed public opinion which can lead to new public policy for wastewater treatment and disposal. In my judgment, the greatest immediate challenge, and opportunity for future program development, lies in this area.

The Third National Conference has brought us to a point of needing to define an overall subdivision wastewater management concept. This new concept must include how the individual onsite system may be incorporated into area development planning, with or without sewers and other treatment/disposal methods.

With appropriate interim preparation, this last stated issue may well be the basis and point of attack for a Fourth National Conference.

INDEX

acquired property disposition 47
advanced wastewater treatment (AWT) 9,38
administration of local programs 67 *ff.*,213,214
aeration units 208,237
aerobic devices/systems 28,29, 45,50,80,85,156
aerobic digestion 160,161
aerobic unit 121-126
 effluent 114-116,121
 pressurized distribution (dosing) 33
air entrainment 177
alternating beds 32,33,53
alternating valves 33
American Society of Civil Engineers 172
anaerobic digestion 45,50,161, 162
Appalachian Project 279
Appalachian Regional Commission 5,235,236,242
attached growth systems 160

Beltsville process 153
Bernhart, A. 99
BIF Purifax process
 See Purifax
biochemical oxygen demand
 See BOD
blackwater 25,29,85,217
BOD 152-154,158-161,278
 removal 173
 standards 108,109

Boyd County (Kentucky) demonstration project 5,235-243, 279
bulking 158

Canadian Standards Association 84
capillary suction time (CST) 163,164
carbon filters 9
carcinogens 7,8
certification of systems 214,216-219
cesspools 78
charcoal 8,9
chemical oxygen demand
 See COD
chlorination 8,120,123,124,153, 154,309
chloroform, formation of 8
civil service status for regulatory officials 61
clarifiers 182
climate 77,95
collection systems 23-25
Colorado Water Quality Control Act 205
Commerce, U.S. Department of 215,221
COD 152,154
 reduction 161-163
composting 45,165
construction
 costs 4,68,69,247,257,268, 282,288-292

315

construction
  effects of   288,293
  restrictions   67,218
costs of sewage systems   13,29, 32,35,38-41,59,107,165,172, 214,236,238,246,248,264,266, 280 ff.,296

data analysis of systems   48
design of sewage systems   44,48, 91,155,165,192,196-200,212, 213,227,232,272,278,279, 285-288,292,301-304,309
Dickerson, Maryland AWT plant   9
disinfection   33,107,109,113-116, 123,180,232,239-241,309,310
disposal fields   285-287
distillation   230
Dow Chemical Company   8
Douglas County, Oregon Special Projects Division   171
drainfields   49,50,131,134,171, 180,181
  alternating   53

effluent pumps   175,176
effluent standards   8,10,38,40, 68,104-111,278,308
electric-thermal subsystems   223-228
electroosmosis   131-135
electroosmotic dewatering   132
elevated temperature biological processes   230
energy resources   68,223-228, 307,308
Energy Research and Development Program (ERDA)   221
environmental impact   68,70,106-111,264,288,293-296
environmental legislation   1-3,10, 12
Environmental Protection Agency (U.S. EPA)   4-6,9,10,21-25, 28,30,38-41,44,147-150,156, 157,172,221,229,240,241,277, 293,298,304

Environmental Protection Agency
  construction grants   35-40, 257,270
  Office of Research and Development   21,25
  pilot plants   22,30
  R&D grants   25
  Technology Transfer Program   41,311
erosion   293
evapotranspiration systems   22, 27-29,32-34,50,66,93-100,181, 199,202,208,237,239,254

failure of sewage systems   27,39, 46,47,59,63,87-90,171,187,213, 218,219,237,238,245,248,256, 270,297
Farmers Home Administration   172,215,241,242,270,271,284
  loans   284
fecal coliform limits   107,125
federal funding
  See funding
Federal Housing Administration (FHA)   215,278
Federal Water Pollution Control Act   249,272
  Amendments   2,3,16,37,247
flow reduction alternatives   34
food/microorganism (F/M) ratio   157-158
Fountain Run, Kentucky   255,277 ff.
funding   37,165,195,248,257,298
  federal   4,22,23,236,237,240-243,248,250,270,280,284,304

General Dynamics Corporation survey   29
gravity sewers   179
Great Lakes-Upper Mississippi River Board (GLUMRB)   201-204
greywater   25,29,30,34,45,78,85, 217,255,303,304
grinder pumps   174,175,253
groundwater   24,25,32,79,88,171, 178,203,206-208,212,294,295

Harkin, J. 90
heavy metal content 30,31,150
holding tanks 80,99,100,239,249
Hoover Plumbing Code 215
Housing and Urban Development, Department of (HUD) 43-51,221,223,227
hydraulic loading 114,115,117-119,122,123,157,176
hydrogen peroxide treatment 33

incineration 228-230
infiltration-percolation ponds 254
infiltration rates 117,119,122
inspection of sewage systems 56,57,214-217
installation of sewage systems 60,61,81,196-199,208,209, 216,217,238-240
interceptor tanks 173-176,181-184
Interior, Department of 227
iodine 241

lagoons 151,154,164,256,257,293
 New York State guidelines 152
land disposal 164,199,229,230, 253,279
land use 16-18,57,77,95,106, 171,184,269
leaching beds 40,78-80,83,84 208,219
Lebo process 153,164
Lederberg, J. 8-9
legislation 15,16,55,75,76,188, 197,205-208
licenses
 See permits

Maine Guidelines for Septic Tank Sludge Disposal on the Land 151
Maine Plumbing Code 215,216
maintenance 238,243
management of sewage systems 189-192,195-200,205-209,215-219,246,258,312
 contracts 190,191

Manual of Septic Tank Practice 32,92,196
mechanical dewatering 163
metals concentrations 149
 Also see heavy metals
MIUS Program 221-233
 cost 222
Modular Integrated Utilities Systems *See* MIUS
monitoring of sewage systems 48,49,57,86,127,162,272,303
mound systems 27,28,32-34,93-100,180,199,219,260

National Aeronautics and Space Administration (NASA) 221
National Bureau of Standards (NBS) 221
National Commission on Water Quality 3
National Drinking Water Advisory Council 8
National Pollutant Discharge Elimination System (NPDES) 207
National Sanitation Foundation (NSF) 50,53,82,85,209,235, 303,304,311,312
 standards 85,303
New York State Department of Environmental Conservation 213
New York State Design Bulletin 211-213
nutrient standards 108

Oak Ridge National Laboratory 221,227
Ontario Environmental Protection Act 75,76
operation/maintenance (O/M) 22, 23-25,59-62,192,227,288
 costs 38-41,126,164,165,230, 246,250,270,283,288
oxidation lagoons 105-111,279

Pennsylvania Department of Environmental Resources 93-98,198

Pennsylvania Sewage Facilities Act
    15,16,19,197
permits   56,57,60,62,81,142,196-
    198,207,213,214,217,218
phosphorus standards   105
PL 92-500   10,21,40,41,45,65,
    66,188,277,283
plumbing permit procedure   218
polishing filter   239
population density   77
present worth analysis   266-268,
    283,296
pressure sewers   21-25,171-184,
    199,228,230,232,252,261,262
public awareness   29,195,297,
    301-305,312
public health   9-11,16,18,40,55,
    59,65,66,78,84,87,103-111,
    188,198,241,245,303,304,312
Public Health Service, U.S.   47,
    196,217
public information   20,100,197
Purifax process (BIF)   152-154,
    164
pyrolysis   229

Randolph, Senator J.   1,236,240
recycling systems   45,162,231,
    232,237-240
regulation of sewage systems
    17-19,58,65,128,213,216,218,
    242,243
regulatory agencies   53-59,65-68,
    87,103-110,127,128,187-189,
    196-198,202,248,288
rejuvenation   90
repair of sewage systems   91,208,
    218,219
research   70,100,201-203,310
reverse osmosis   230
rotating biological contactors
    160
rural sewage systems   4,16-19,
    21 ff.,37 ff.,172,184,198,199,
    235-243,255

Safe Drinking Water Advisory
    Council   8

sand filters   33,79-80,85,86,105-
    111,114,119-124,180,181,219,
    254,261,262,280,281,295
sand drying beds   154,155,163,
    164
sand mounds
    See mounds
Seattle's Metro Renton facility
    156
Senate Public Works Committee
    1-5
septage   22,23,30,31,114-120,
    147 ff.,156-165
septic tanks   30,33,47,49,64,78-
    80,84,87-90,131,147-150,165,
    173-175,180-184,208,219,236,
    248,252,255,260,265,268,270,
    287,288,294
septic tank-soil absorption systems
    (ST-SAS) 25-27,32-34,54,113,
    127,253
service lines   176
Sewage Osmosis Systems   131-135
sewer mains   176,177
site selection   56,59,94,95,139-144
sludge
    See septage
slurry systems   228
small-diameter gravity sewers
    252,261-265,298
Small Flows Research Program
    22,23,30,35
Small-Scale Waste Management
    Project   96,113,128,258,272
soils   47,50,76,79,88-91,94,95,
    113,131-135,139-144,203,215-
    217,248,252,256,280,281,296,
    309,310
soil absorption system (SAS)   27,
    139-144,254,261,262,265,270,
    280
soil crust test   143
soil maps   143,144
soil percolation   33,78,80-83,88,
    132,133,142-144,196,206-208,
    215,216
    tests   82,83,88,196,215,216
soil sampling tube   88
split-stream systems   303
spray irrigation   151,199,219,237

standards for sewage systems 55, 60,74,81,82,100,104,196-198, 201-203,211-214,242
Stokinger, H. 8
stream-bank damage 293
stream discharge
See surface discharge
subsurface sewage disposal 83, 84,94,173,181,197-199,215, 237,238,250,280,285 *ff.*,297, 298
Supreme Court, U.S. 56,57
surface discharge 26-28,32,34, 65,103,113,127,128,171,172, 180,196,198,218,237,238,250, 253,254,261,262,278
suspended solids 108,109,173
System Approach to Individual Home Treatment 235-243

Ten State Committee for Onsite Sewage Systems 197, 201-204
terrain 76,77
tertiary systems 307
Toronto, University of 99
toxic substances 8,10
  industry responsibility 10
Toxic Substances Control Act 10
trace elements 7,9
trench alternatives 34
trickling filters 160
toilets, types of 78,218
  biological 34
  chemical 46
  composting 85
  electric 46
  incinerator 34
  low-flush 34
  waterless 29,34

ultraviolet irradiation 125,126
  cost of 125,126
  units 116
Underwriters Laboratories 84
Universal Soil-Loss Equation 293

utility systems 222-227,233

vacuum filters 163
vacuum sewage systems 22-25, 199,228,230,232,253
Virginia Polytechnical Institute 100
VSS 152,161,162
  reduction 161,162

waiver procedure 218,219
Washington State Health Department 47
Washington, University of 48,49
  pilot project, retrofitting of drainfields 45
waste disposal options 75 *ff.*
waste recycling 12,13
waste segregation 254
wastewater analyses 50
wastewater characteristics 115
wastewater composition 25,26
wastewater generation 25,26
wastewater health effects 8
wastewater recycling 5,7,30
wastewater systems
  complaint analysis 72
water conservation 35,85,254
Water Pollution Control Federation 172
water quality objectives 114
water quality standards 7,8,65,85
water resources 7
water tables 83,88,89,151
  Also see groundwater
waterless toilets 29,34
Weaverville Waste Water Treatment Plant 157
Wisconsin Department of Natural Resources 270
Wisconsin, University of 22,25,27, 28,90,96,116,180

zero discharge 7,11,12
zoning 58